Novel Cementitious Materials

Novel Cementitious Materials

Topic Editors

Peiyu Yan
Yao Luan
Chunsheng Zhou

MDPI

Basel • Beijing • Wuhan • Barcelona • Belgrade • Novi Sad • Cluj • Manchester

Topic Editors

Peiyu Yan
Department of Civil Engineering
Tsinghua University
Beijing
China

Yao Luan
Department of Civil and Environmental Engineering
Saitama University
Saitama
Japan

Chunsheng Zhou
School of Civil Engineering
Harbin Institute of Technology
Harbin
China

Editorial Office
MDPI AG
Grosspeteranlage 5
4052 Basel, Switzerland

This is a reprint of the Topic, published open access by the journals *Buildings* (ISSN 2075-5309), *Materials* (ISSN 1996-1944), and *Solids* (ISSN 2673-6497), freely accessible at: https://www.mdpi.com/topics/7PVP0TWV56.

For citation purposes, cite each article independently as indicated on the article page online and as indicated below:

Lastname, A.A.; Lastname, B.B. Article Title. *Journal Name* **Year**, *Volume Number*, Page Range.

ISBN 978-3-7258-6962-6 (Hbk)
ISBN 978-3-7258-6963-3 (PDF)
https://doi.org/10.3390/books978-3-7258-6963-3

Contents

Preface

Carbon neutrality and peak carbon dioxide emissions are hot topics to ensure the Earth's sustainable development. Cementitious materials are the most commonly used artificial engineering materials worldwide. The production of Portland cement contributes a considerable portion of carbon dioxide emissions. Some novel composite cementitious materials containing various kinds of solid waste were developed to reduce the carbon footprint of concrete. For instance, novel cements with low carbon dioxide emissions were studied and began to be used in different industries, such as UHPC, ECC, geopolymers, some non-Portland cements, etc. They displayed many interesting performances that satisfied the increasing requirement for infrastructure with mechanical and durable properties, with a heavy load in severe built environments. Their hydration mechanism and properties are not the same as those of typical Portland cement. The relationship of their composition, microstructure, and properties has been comprehensively investigated in recent decades. This Reprint includes articles focused on recent developments related to the preparation and microstructural characteristics of novel cementitious materials, as well as their properties and applications in different fields. We hope that this Reprint will be very useful for readers to more deeply understand novel cementitious materials.

Peiyu Yan, Yao Luan, and Chunsheng Zhou
Topic Editors

materials

MDPI

Article

Formation, Stability, and Crystallinity of Various Tricalcium Aluminate Polymorphs

Simona Ravaszová [1], Karel Dvořák [1,*], Martin Boháč [2], Dalibor Všianský [3] and Andrea Jančíková [1]

1 Faculty of Civil Engineering, Brno University of Technology, Veveří 331/95, 602 00 Brno, Czech Republic; ravaszova.s@fce.vutbr.cz (S.R.); janciku.a@fce.vutbr.cz (A.J.)

2 Research Institute for Building Materials, Hněvkovského 30/65, 617 00 Brno, Czech Republic; bohac@vush.cz

3 Department of Geological Sciences, Faculty of Science, Masaryk University, Kotlářská 267/2, 611 37 Brno, Czech Republic; dalibor@sci.muni.cz

* Correspondence: dvorak.k@fce.vutbr.cz; Tel.: +420-775906555

Abstract: Tricalcium aluminate is an important phase of Portland clinker. In this paper, three polymorphs of C_3A were prepared by means of the solid-state synthesis method using intensive milling of the raw material mixture which was doped with various amounts of Na_2O and sintered at a temperature of 1300 °C for 2 h. The final products were evaluated through X-ray diffraction using Rietveld analysis. The effect of the Na dopant content on the change in the crystalline structure of tricalcium aluminate was studied. It was proven that the given preparation procedure, which differed from other studies, was close to the real conditions of the formation of Portland clinker, and it was possible to prepare a mixture of different polymorphs of calcium aluminate. Fundamental changes in the crystal structure occurred in the range of 3–4% Na, when the cubic structure changes to orthorhombic. At a dosage of Na dopant above 4%, the orthorhombic structure changes to a monoclinic structure. There are no clearly defined boundaries for the existence of individual C_3A phases; these phases arise at the same time and overlap each other in the areas of their formation at different Na doses.

Keywords: tricalcium aluminate; polymorphism; cubic; orthorhombic; monoclinic; crystallite size

1. Introduction

Due to economic and environmental pressures, especially regarding the production of a lower volume of CO_2, modern cement materials are increasingly complex from a chemical and mineralogical point of view [1]. The use of alternative fuels such as waste materials, biomass, and municipal solid waste can reduce our dependence on fossil fuels, reduce greenhouse gas emissions, and improve energy efficiency. However, this represents not only economic advantages but also technological challenges. This trend inevitably leads to a change in the properties of the individual phases of Portland cement. The properties of these components can therefore differ significantly from those expected for traditional Portland cement in previous years. This requires extensive research and testing [2–6].

Tricalcium aluminate (C_3A) is one of the most important components of Portland cement (OPC). Tricalcium aluminate with a melting point of 1540 °C belongs to the group of the most represented calcium–aluminate phases in clinker, forming approximately 7 to 15 wt.% of it [7,8]. Currently, the formation of tricalcium aluminate during sintering is often affected by various foreign ions. These ions change the resulting structure of tricalcium aluminate and thus also the course of hydration of the cement phase. The most common C_3A modifiers are sodium and potassium ions in the form of alkalis [9]. They tend to be contained in larger or smaller quantities in raw materials to produce clinker. Their content is measured and controlled, which enables additional correction. In contrast, monitoring the quantitative representation of ions introduced into the material due to the burning of solid alternative fuels is relatively difficult to implement. Nowadays, cement plants use a whole range of secondary fuels for firing clinker in the production of cement. The most commonly used secondary raw materials in incineration are waste oil, used tires, meat, and bone meal or sorted waste, which includes mixed plastics, textiles, textile fibers, carpets, rubber, paper, plastic–paper composite packaging, and chipboard [7–10].

The most important property of C_3A affects the processability of cement paste [11,12].

C_3A occurs in clinker in three modifications depending on the content of alkalis, especially sodium ions. Other oxides that cause C_3A modification are K_2O, SiO_2, MgO, and Fe_2O_3 [7,8,13,14]. Pure tricalcium aluminate occurs in a cubic structure. Modification with Na_2O in particular changes the cubic structure to an orthorhombic structure, monoclinic structure, and two metastable structures, i.e., tetragonal (high-temperature polymorph) and a weakly crystalline form, the so-called "proto-C_3A". The reactivity of the given structures is different and so is their course of hydration [15–17].

One calcium ion is replaced by two sodium ions. One sodium ion takes the place of the original Ca^{2+} and the other is in the center of the $(Al_6O_{18})^{18-}$ ring. The general formula of C_3A modified by sodium ions can be written as $Na_{2x}Ca_{3-x}Al_2O_6$ where x denotes the amount of calcium ions substituted by sodium ions. As x increases from 0 to 0.25, the crystal system changes from cubic to orthorhombic and then to monoclinic at room temperature. In the C_3A crystal structure, there are eight sites on the triaxial axes where additional atoms can be inserted. The size of Ca^{2+} and Na^+ ions is similar, and it is therefore possible to state that Na^+ ions replace Ca^{2+} atoms in their places in the structure [18,19]. Sodium (Na), which replaces calcium (Ca) in the crystal lattice, occupies a space that is about the same size but slightly larger. When calcium is substituted for sodium, the gaps between the atoms are not filled, leading to a gradual change in the crystal structure. As a result, the initially symmetric cubic lattice becomes a less symmetric orthorhombic or monoclinic [20,21].

Tricalcium aluminate is formed by a solid–solid reaction between CaO and Al_2O_3 indirectly through the phases, with various ratios between CaO and Al_2O_3. Despite long-term application, there is still no consensus regarding the kinetic and mechanistic aspects of its formation [22–24].

The preparation of high-purity C_3A single phases differs considerably from industrial production. In addition to the classic method based on a high-temperature reaction in the solid phase, there are also alternative methods such as the sol–gel method, combustion syntheses, or synthesis using mechanochemical activation.

The literature [25] states that mechanochemical activation can significantly affect the rate of reactions, whereby the most reproducible and consistent results were obtained under conditions of good interparticle contact with controlled pretreatment to define the physical structure. Mechanochemical activation supports the course of reactions in the solid phase, as it increases the specific surface area and local thermal effects. This is a very energy-intensive preparation, in which the individual processes occur at lower temperatures, however, which, on the other hand, makes it possible to reduce the overall temperature and firing time. The course of the reaction is similar to the high-temperature method [26,27].

Finding a suitable procedure for the sol–gel synthesis of calcium–aluminate phases turned out to be somewhat problematic. For the preparation of calcium–aluminate phases, the Pechini method and its modified versions, which also belong to the group of sol–gel methods, have proven themselves. However, these methods are very far from the real conditions for the formation of calcium aluminate phases in Portland clinker [28].

Therefore, this article is devoted to the traditional method of reaction in the solid phase, which represents the oldest method used for the preparation of the calcium–aluminate phase, but it is still the most frequently used nowadays. In this method, thorough homogenization of the raw materials is very important. If the starting mixture is not homogeneous, the product tends to be contaminated with unreacted starting substances. Preparing a pure phase of the desired composition is often problematic with this method. Unwanted phases are often present in the product. They are present in the product if the reaction does not take place completely or if there are inhomogeneities in the reaction mixture. In the area of inhomogeneities, local phase equilibrium is established, and the corresponding phase is formed.

Previous studies [9,15,16,19–21,29–31] on calcium–aluminate phases often use a different methodology than the one used in this study. This article is devoted to the preparation of various C_3A polymorphs when the raw material mixture is doped with sodium, which is added to the raw material mixture before sintering. The aim is to study the formation of individual polymorphs that are formed in parallel. In this way, the preparation procedure presented in this study is close to the real conditions of Portland clinker production. The goal is to prepare pure phases through solid-state synthesis in the highest possible purity.

This article deals with the preparation of various polymorphs of tricalcium aluminate by means of the solid-state method. The aim is to use a simple preparation method that includes mechanical activation of the raw material mixture doped with different Na content and one-step firing to prepare tricalcium aluminate at the highest possible purity.

To evaluate the purity of the prepared C_3A samples, the Rietveld method was chosen, which involved the Sherrer equation to monitor the crystallinity of the prepared tricalcium aluminates.

2. Materials and Methods

2.1. Raw Material Mixture Composition

Calcium carbonate ($CaCO_3$, p.a. purity, Penta, Praha, Czech Republic), aluminum oxide (Al_2O_3, p.a. purity, Penta, Praha, Czech Republic), and sodium carbonate (Na_2CO_3, p.a. purity, Penta, Praha, Czech Republic) were used as precursors for the synthesis of tricalcium aluminate. The composition of the raw material mixture for the preparation of tricalcium aluminate was based on already published data [9,30]. The $CaCO_3$ and Al_2O_3 were mixed in a stoichiometric ratio of 3 mol CaO and 1 mol Al_2O_3, which corresponds to the cubic C_3A modification. Stabilization with sodium (added in the form of sodium carbonate) is necessary to obtain the orthorhombic and monoclinic forms. The content of Na_2O added to the raw material mixture for the formation of orthorhombic and monoclinic C_3A is shown in Table 1.

Table 1. Composition of Na-substituted C_3A raw material mixture.

Label	$CaCO_3$ [g]	Al_2O_3 [g]	Na_2CO_3 [g]	Total
0.0% Na_2O	22.40	7.60	0.00	30.000
0.5% Na_2O	22.40	7.60	0.26	30.260
1.0% Na_2O	22.40	7.60	0.51	30.510
1.5% Na_2O	22.40	7.60	0.77	30.770
2.0% Na_2O	22.40	7.60	1.03	31.030
2.5% Na_2O	22.40	7.60	1.28	31.280
3.0% Na_2O	22.40	7.60	1.54	31.540
3.5% Na_2O	22.40	7.60	1.8	31.800
4.0% Na_2O	22.40	7.60	2.05	32.050
4.5% Na_2O	22.40	7.60	2.31	32.310
5.0% Na_2O	22.40	7.60	2.57	32.570
5.5% Na_2O	22.40	7.60	2.82	32.820
6.0% Na_2O	22.395	7.605	3.08	33.080

In order to ensure a sufficient amount of Na, which suffers from the relatively high volatility of sodium at high temperatures and resulting losses during sintering, Na_2CO_3 was added to the calculated stoichiometric ratio of CaO and Al_2O_3.

2.2. Tricalcium Aluminate Preparation

Milling and homogenization of the raw material mixture were carried out through high-energy wet milling in a planetary mono mill (Pulverisette 6, Fritsch, Idar-Oberstein, Germany). A 40 mL steel capsule and 25 steel balls with a diameter of 20 mm were used. The basic ratio of dry ingredients and water was 30 g of the raw material mixture and 30 g of water. The water/powder ratio of 1.0 was used. The milling mode of 350 rpm for 15 min was selected. A Malvern Mastersizer 2000 laser granulometer (Malvern Panalytical Ltd., Malvern, UK) with a hydro 2000 G fluid dispersion unit, using 2- 2-isopropanol as the dispersing agent was used for determining the granulometry. After wet milling, all samples were dried in a laboratory oven (Binder C170, Binder GmbH., Tuttlingen, Germany) at a temperature of 105 °C for 24 h. The slurry spontaneously formed solid nodules during the drying process, which were subsequently placed in 30 mL platinum crucibles. The sintering process was carried out in a high-temperature furnace (2017 S, Clasic CZ s.r.o., Řevnice, Czech Republic). The firing mode was designed so that the carbonates first decompose at a temperature of 1000 °C for 1 h and then, the synthesis itself takes place in the solid phase at a temperature of 1300 °C for 2 h. The heating rate was 8 °C/min. At the end of the firing mode, the crucibles were removed from the furnace and immediately cooled by a stream of cold air to room temperature. The sintered products were subjected to milling in a vibratory disc mill (RS 200, Retsch, Haan, Germany) at 1000 rpm for 20 s. After that, 5 g of sintered products in the form of powder was milled to the required fineness for X-ray analysis in a mill (McCrone Micronising Mill, Glen Creston, London, UK) for 150 s in 15 mL of isopropanol.

2.3. X-ray Diffraction Analysis

The XRD analysis was performed using a multifunctional diffractometer (XRD, Empyrean, PANalytical B.V., Almelo, The Netherlands). The Θ-Θ reflection Bragg–Brentano para focusing geometry device is equipped with a Cu anode (λ = 1.54184 Å), programmable divergence slits, and a PIXcel3D detector(Empyrean, PANalytical B.V., Almelo, The Netherlands) with 255 active channels, with an accelerating voltage of 45 kV, a beam current of 40 mA, and a diffraction angle of 2 Θ in the range from 5 to 60° with a step size of 0.013° and 38 s per step. The total time per scan was 12 min. Each sample was measured four times, and the scans were next simply summed. The polymorphic purity of the samples was verified via XRD using the Rietveld method. The ICSD database (released in 2012) was used to qualitatively analyze the diffraction patterns. HighScore plus software (3.0e, PANalytical B.V., Almelo, The Netherlands) was used to identify the individual phases, perform their quantification, and determine the amount of the amorphous phase.

2.4. Tricalcium Aluminate Crystallinity

The size of the crystallites was evaluated on a selected diffraction line with the major crystallographic direction. Calculation of the crystallite size is based on the measurement of full width at half maximum (FWHM) via the Scherrer equation [32] with Warren correction [33]; see Equation (1):

$$L = \frac{K \cdot \lambda}{\cos\theta} \cdot \frac{1}{\beta} = \frac{K \cdot \lambda}{\cos\beta} \cdot \frac{1}{\sqrt{B^2 - b^2}} \tag{1}$$

L is the crystallite size, K is the Scherrer constant (the value of 0.89 was used), $\lambda = K_{\alpha 1}$ is the wavelength of X-ray radiation, θ is the diffraction angle, B is the FWHM, and b is the FWHM of the size/strain standard used (lanthanum hexaboride).

The instrument broadening was defined using the LaB_6 standard. LaB_6 peaks were determined in the positions identical to the selected diffraction lines using HighScore plus software, see Figure 1. The FWHM values of LaB_6 were calculated from Equation (1) [34].

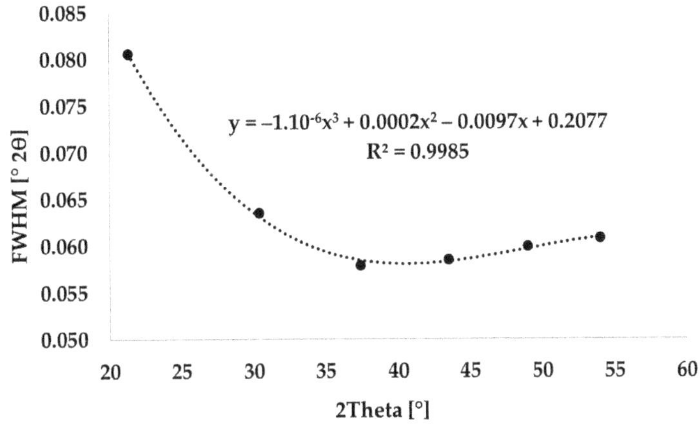

Figure 1. Instrumental broadening measured using LaB_6.

3. Results and Discussion

3.1. Tricalcium Aluminate Preparation

The distribution of particles after the homogenization of raw materials through high-energy milling was determined using laser granulometry; see Figure 2.

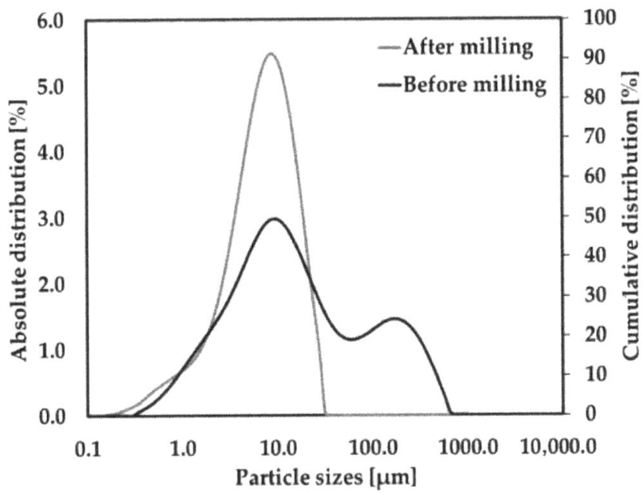

Figure 2. Particle distribution (PSD) of the raw material mixture.

The median particle size of the raw material mixture before homogenization in the Pulverisette 6 laboratory mill (Pulverisette 6, Fritsch, Idar-Oberstein, Germany) was 12.3 μm. After 15 min of milling at 350 rpm, the raw material mixture was refined with a median particle size of 6.8 μm.

3.2. X-ray Diffraction Analysis

Table 2 shows the quantitative mineral composition of each sample. The results of the Rietveld analysis are presented in Appendix A.

Table 2. Rietveld refinement quantified phase composition of the samples.

Na_2O [%]	C_3A-c	C_3A-o	C_3A-m	Mayenite	f-CaO	Portlandite	Total	GOF
0.0	89.7	0.0	0.0	0.0	8.0	2.3	100.0	2.55
0.5	94.4	0.0	0.0	0.0	4.5	1.2	100.1	2.48
1.0	92.1	0.0	0.0	2.7	3.2	2.0	100.0	2.54
1.5	90.5	0.0	0.0	3.1	4.7	1.6	99.9	2.49
2.0	88.8	0.0	0.0	4.4	4.6	2.2	100.0	2.59
2.5	89.9	0.0	0.0	4.0	4.0	2.2	100.1	2.60
3.0	66.9	24.3	0.0	4.0	0.8	4.0	100.0	2.30
3.5	52.9	36.8	0.0	5.0	0.6	4.6	99.9	2.48
4.0	22.2	12.2	48.9	4.8	0.6	11.3	100.0	1.69
4.5	14.3	12.0	58.3	3.8	0.7	10.9	100.0	1.68
5.0	11.0	5.6	69.5	4.0	0.4	9.5	100.0	2.27
5.5	15.7	9.2	61.3	3.0	0.5	10.2	99.9	2.19
6.0	9.5	8.4	67.0	4.7	0.7	9.6	99.9	2.47

The majority of C_3A phases were identified on the diffraction patterns: C_3A cubic (ICSD 00-038-1429), C_3A orthorhombic (ICSD 1880), and C_3A monoclinic (ICSD 100221). The mineral mayenite (ICSD 00-009-0413), lime (ICSD 00-043-1001), and portlandite (ICSD 00-004-0733) as minor phases were also identified.

Figure 3 graphically shows the effect of Na doses on the change in the C_3A polymorphism. The presence of higher Na content is a basic prerequisite for the formation of non-cubic C_3A phases. It can be seen in Table 2 and Figure 3 that at least 2.5–3% Na is needed for the formation of non-cubic phases of tricalcium aluminate. At Na content of 3 to 4%, the formation of the orthorhombic C_3A is supported. At higher doses of Na above 3.5–4%, the monoclinic phase C_3A is formed.

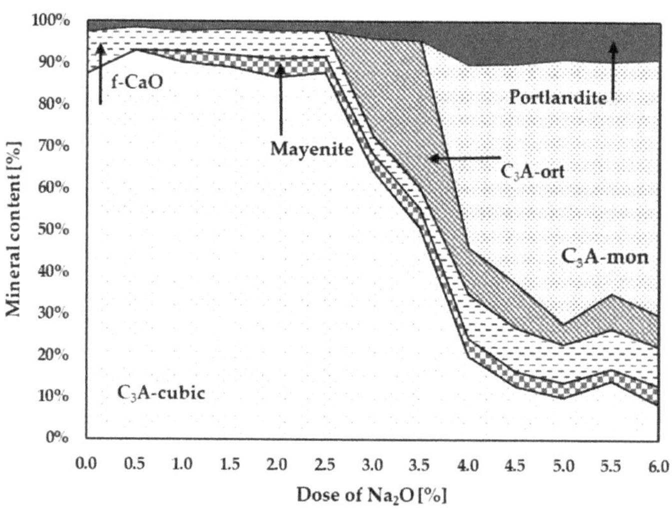

Figure 3. Graphical representation of individual mineral contents at different Na contents.

The presence of portlandite is caused by the very rapid hydration of lime due to air moisture. The reason for the higher content of lime, i.e., portlandite, can be determined according to another study [31], with the release of free lime during the formation of a solid solution of C_3A with sodium ions calculated according to the following equation:

$$3Ca_3Al_2O_6 + Na_2O \leftrightarrow Na_2Ca_8Al_6O_{18} + CaO \quad (2)$$

The synthesized cubic, Na-substituted orthorhombic, and monoclinic C_3As were evaluated as mixtures, indicating that accurate quantification is very difficult to perform.

Identification and quantification of cubic C_3A is generally easy because the diffraction lines are well defined and clearly match the reference pattern. Substitution of Na, however, causes significant difficulties because the pattern of cubic and orthorhombic or monoclinic forms significantly overlaps, see Figure 4. Therefore, it is quite difficult to clearly and accurately fit the individual diffraction lines of cubic, orthorhombic, and monoclinic C_3A using the XRD technique.

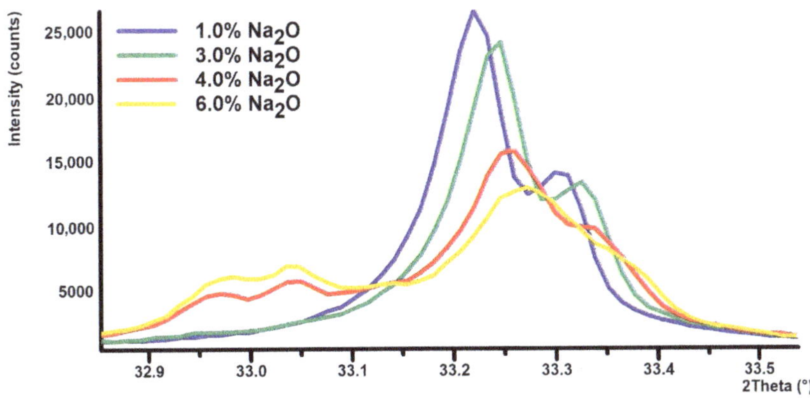

Figure 4. Portion of XRD powder patterns: 1.0% Na_2O (cubic C_3A), 3.0% Na_2O (cubic + orthorhombic C_3A), 4% Na_2O (cubic + orthorhombic + monoclinic C_3A), and 4% Na_2O (mainly monoclinic C_3A).

Unit Cell Volume

The Rietveld method was also used in studying the crystal structure to provide more detailed information about the crystal lattice of the individual polymorph phases of tricalcium aluminate. Table 3 shows the unit cell volumes determined using the Rietveld method, graphically in Figure 5.

Table 3. Unit cell volume of cubic, orthorhombic, and monoclinic C_3A.

Na_2O [%]		Volume [Å^3]		Volume [Å^3]		Volume [Å^3]
0.0		3551.29		0.00		0.00
0.5		3549.51		0.00		0.00
1.0		3546.20		0.00		0.00
1.5		3542.38		0.00		0.00
2.0		3541.13		0.00		0.00
2.5		3540.49		0.00		0.00
3.0	C_3A-cub	3539.06	C_3A-ort	1778.56	C_3A-m	0.00
3.5		3539.69		1779.86		0.00
4.0		3539.58		1786.13		1780.42
4.5		3541.50		1785.70		1779.76
5.0		3540.52		1787.93		1779.63
5.5		3537.95		1786.49		1782.00
6.0		3537.86		1786.68		1788.58

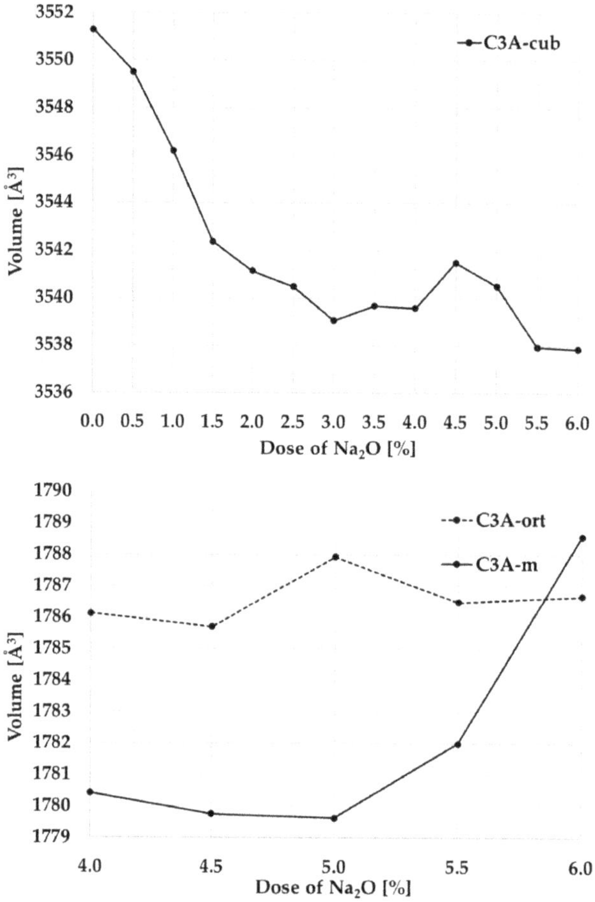

Figure 5. Graphical representation of changes in the unit cell volume at different Na contents.

Sodium (Na), which replaces calcium (Ca) in the crystal lattice, occupies a space that is slightly larger. When sodium is replaced by calcium, the spaces between the atoms are not filled, leading to a gradual change in the crystal structure. As a result, the initially symmetric cubic lattice becomes a less symmetric orthorhombic and subsequently monoclinic structure. At Na content of 3%, the mixture contains cubic C_3A with a unit cell volume of around 3545 $Å^3$ and orthorhombic C_3A with almost half the volume of the unit cell.

3.3. *Tricalcium Aluminate Crystallinity*

To supplement more detailed information and connections with the origin and development of the C_3A polymorphism, the development of crystallinity was monitored.

The size of the crystallites was evaluated in the normal to appropriate crystallographic plane that had the main crystallographic direction and was sufficiently intense. A distinct crystallographic plane determined by the Miller indices "hkl": 004 was chosen, which appears on the diffraction pattern at the position 2θ: 33.177°; see Figure 6.

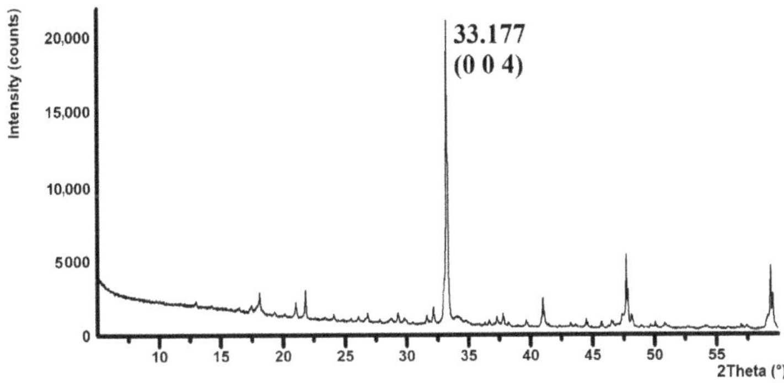

Figure 6. Details of the reflection of the selected crystallographic plane (0 0 4) on the XRD pattern.

The Sherrer method was used to calculate the crystallite size. Table 4 and Figure 7 graphically show the crystallite sizes for individual samples.

Table 4. Crystallite size of cubic, orthorhombic, and monoclinic C_3A (L Vol-FWHM).

Na_2O [%]		L [Å]		L [Å]		L [Å]
0.0		1233.6		0.0		0.0
0.5		1184.3		0.0		0.0
1.0		1264.8		0.0		0.0
1.5		1233.6		0.0		0.0
2.0		1249.0		0.0		0.0
2.5		1208.4		0.0		0.0
3.0	C_3A-cub	1320.5	C_3A-ort	332.7	C_3A-m	0.0
3.5		1320.5		466.2		0.0
4.0		1320.5		602.2		680.0
4.5		659.3		467.9		1320.5
5.0		457.2		580.5		1047.8
5.5		455.6		711.2		1262.5
6.0		467.4		467.4		1212.5

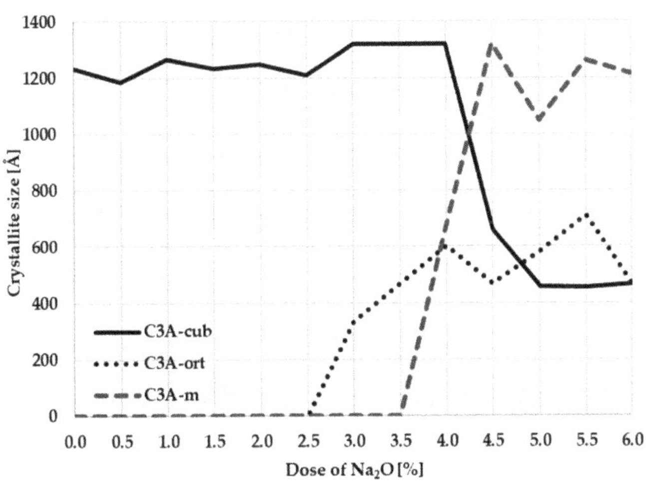

Figure 7. Dependence of crystallite size on Na content.

The course of the crystallite size corresponds to the changes in the unit cell during the transformations of the individual crystal structures. Based on the calculation of the size of the crystallites, it can be stated that the most significant changes in crystallinity occur at Na_2O content of between 3 and 4 percent when there is a fundamental polymorphic change from cubic to orthorhombic. The average crystallite size of cubic C_3A is around 1200 Å. When the sodium ions are incorporated into the C_3A cubic structure, the size of its crystallites rapidly decreases to an average value of 466 Å. At 4% sodium content, the orthorhombic structure changes to a monoclinic structure. The average crystallite size of the monoclinic structure is around 1104 Å and approaches the crystallite size of the original cubic C_3A.

4. Conclusions

The results obtained in this study allow the following conclusions to be reached:

- The various polymorphs of C_3A were prepared through the method of solid-state synthesis and a procedure close to the real conditions of the formation of Portland clinker.
- The presence of multiple polymorphs might be due to the particle size of the raw meal.
- Although powder diffraction has a fundamental limitation in the reliable identification and quantification of C_3A polymorphs, it is the only accessible approach.
- The stability of cubic tricalcium aluminate is up to 2.5% Na content.
- The formation of orthorhombic tricalcium aluminate using solid-state synthesis, as described, requires Na content of at least 3–3.5% Na.
- The transition from an orthorhombic structure to a monoclinic structure takes place at 4–4.5% Na content.
- The most significant changes in crystallinity occur at 3–4% Na content.
- There are no clearly defined boundaries for the existence of individual C_3A phases, these phases arise at the same time and overlap each other in the areas of their formation at different Na doses.

Author Contributions: Conceptualization, S.R. and K.D.; methodology, S.R.; software, K.D. and D.V.; validation, S.R., K.D. and A.J.; formal analysis, M.B.; investigation, S.R. and K.D.; resources, S.R.; data curation, K.D.; writing—original draft preparation, S.R.; writing—review and editing, S.R. and K.D.; visualization, S.R.; supervision, K.D. and S.R.; project administration, S.R.; K.D. and M.B.; funding acquisition, K.D. and S.R. All authors have read and agreed to the published version of the manuscript.

Funding: We acknowledge the financial support given to the project "Study of the formation and stability of various tricalcium aluminate polymorphs" (FAST-S-23-8246) and the project of the Czech Science Foundation "The role combination of fluxes and mineralizers on properties of low-energy clinker" (23-05122S).

Institutional Review Board Statement: Not applicable.

Informed Consent Statement: Not applicable.

Data Availability Statement: Data are contained within the article.

Conflicts of Interest: The authors declare no conflicts of interest.

Appendix A

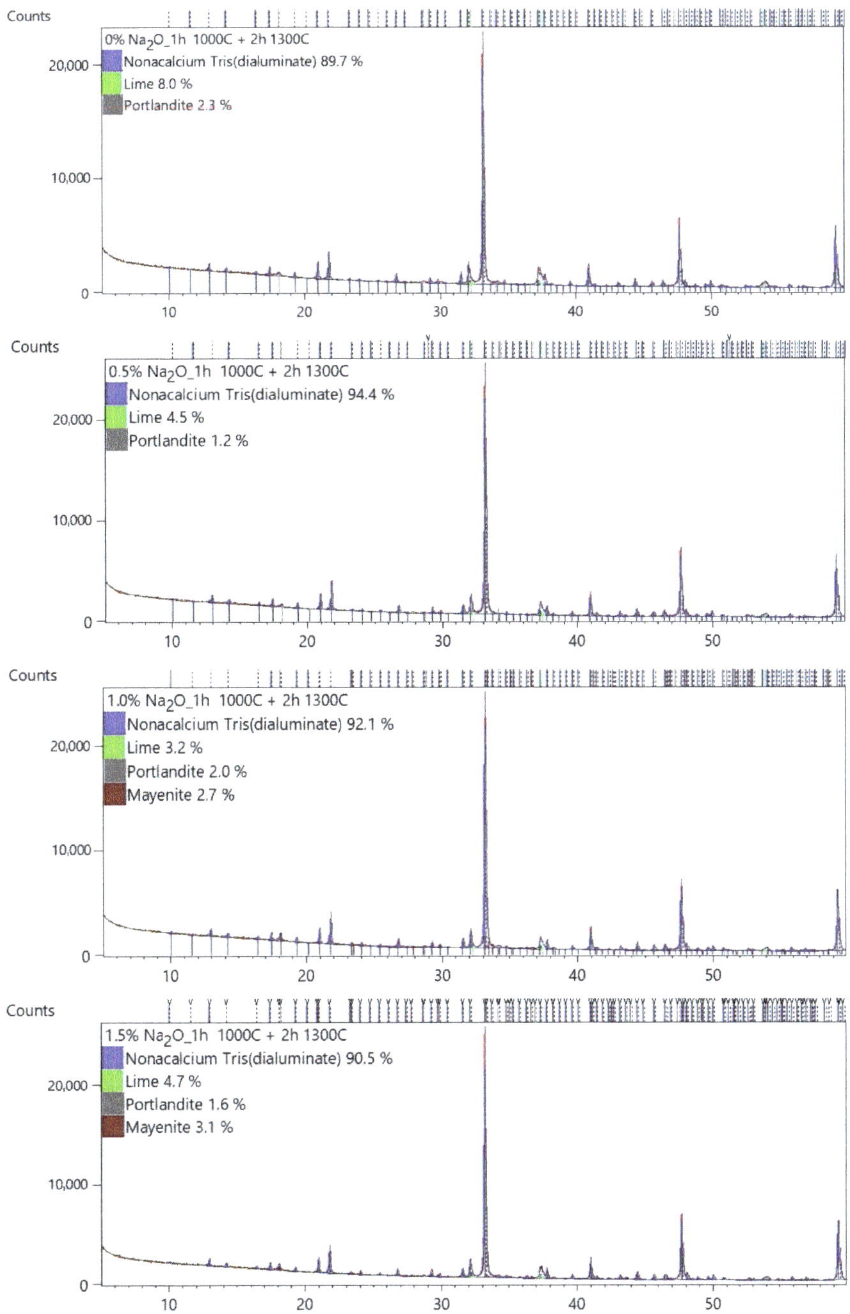

Figure A1. *Cont.*

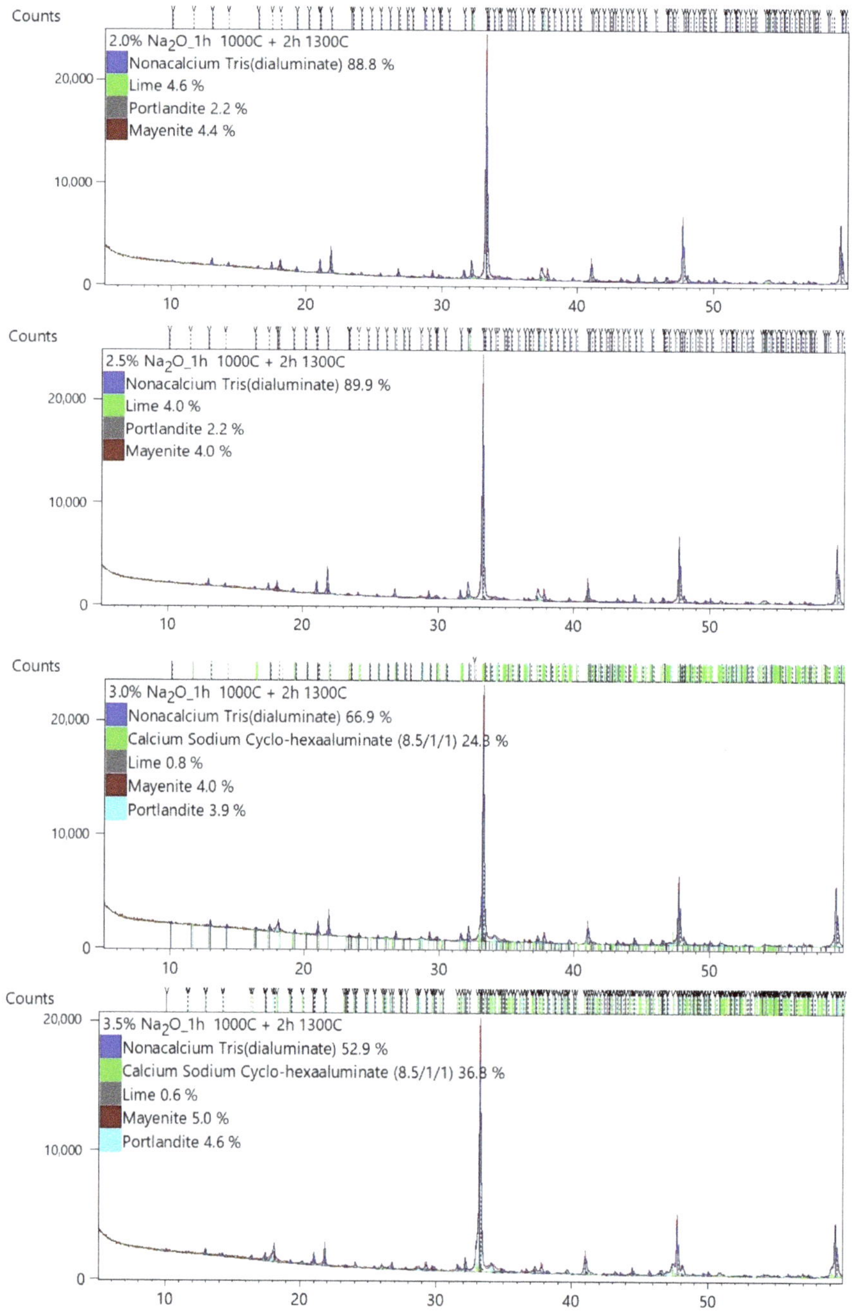

Figure A1. *Cont.*

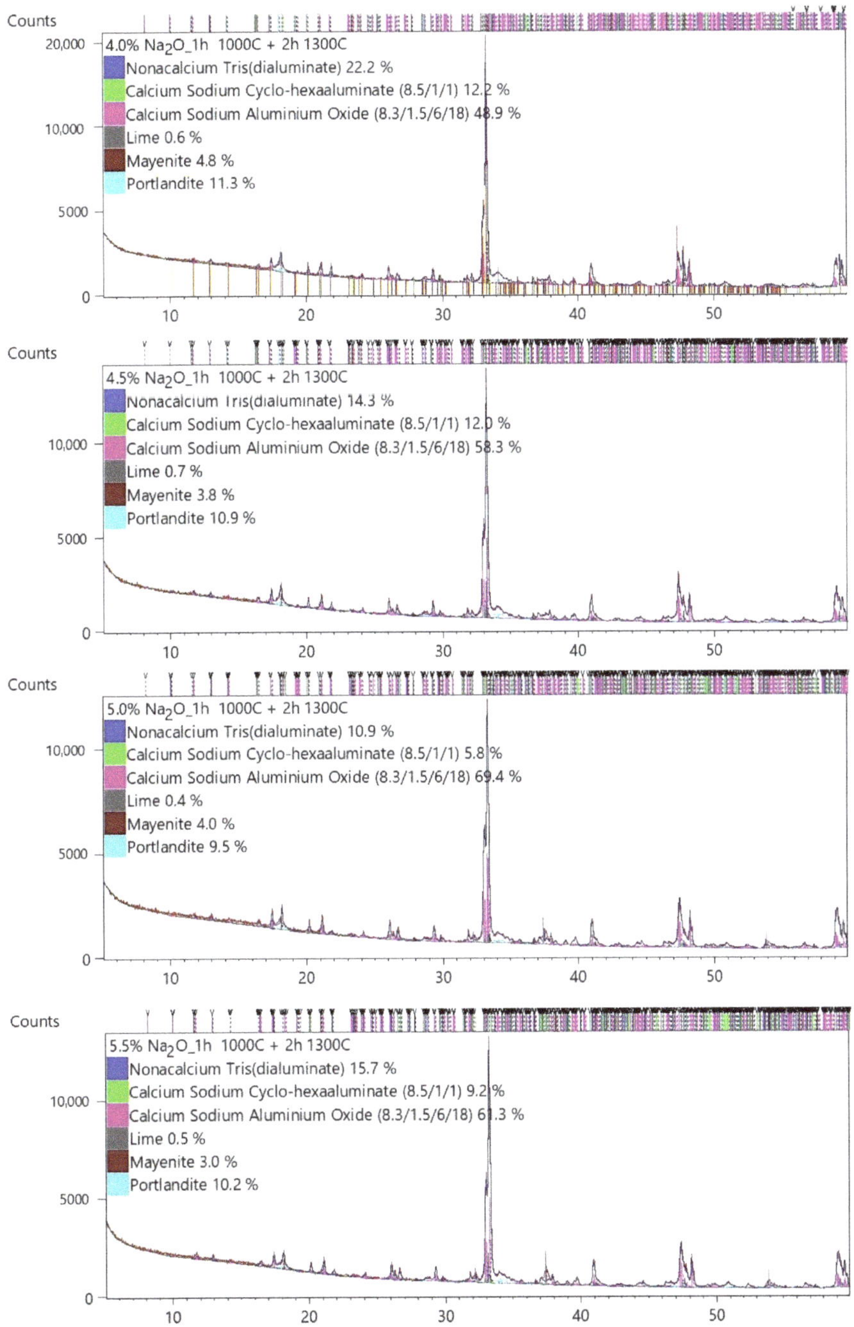

Figure A1. *Cont.*

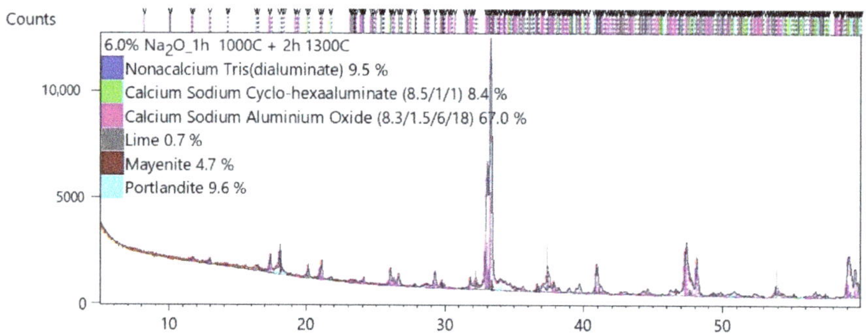

Figure A1. Rietveld analysis results.

References

1. Scrivener, K.L.; John, V.M.; Gartner, E.M. Eco-efficient cements: Potential economically viable solutions for a low-CO_2 cement-based materials industry. *Cem. Concr. Res.* **2018**, *114*, 2–26. [CrossRef]
2. CEMBUREAU, 2050 Carbon Neutrality Roadmap. Available online: https://cembureau.eu/library/reports/2050-carbon-neutrality-roadmap/ (accessed on 29 November 2023).
3. Schneider, M.; Hoenig, V.; Ruppert, J.; Rickert, J. The cement plant of tomorrow. *Cem. Concr. Res.* **2023**, *173*, 107290. [CrossRef]
4. CEMBUREAU, Energy Assessment for CCUS Options EU Cement Sector. Available online: https://cembureau.eu/media/bucbkisd/230503-tno-r10586-1-energy-assessment-for-ccus-options-eu-cement-sector-final.pdf (accessed on 29 November 2023).
5. Szabó, L.; Hidalgo, I.; Císcar, J.C.; Soria, A.; Russ, P. *Energy Consumption and CO_2 Emissions from the World Cement Industry*; European Commission Joint Research Center: Brussels, Belgium, 2003.
6. Mokrzycki, E.; Uliasz-Bocheńczyk, A. Alternative fuels for the cement industry. *Appl. Energy* **2003**, *74*, 95–100. [CrossRef]
7. Taylor, H.F.W. *Cement Chemistry*; Academic Press: London, UK, 1990.
8. Hewlett, P.C. *Leas 's Chemistry of Cement and Concrete*; Arnold: London, UK, 1998; pp. 131–193.
9. Wesselsky, A.; Jensen, O.M. Synthesis of pure Portland cement phases. *Cem. Concr. Res.* **2009**, *39*, 973–980. [CrossRef]
10. Rodríguez, N.H.; Martínez-Ramírez, S.; Blanco-Varela, M.T.; Donatello, S.; Guillem, M.; Puig, J.; Fos, C.; Larrotcha, E.; Flores, J. The effect of using thermally dried sewage sludge as an alternative fuel on Portland cement clinker production. *J. Clean. Prod.* **2013**, *52*, 94–102. [CrossRef]
11. Nakomcic-Smaragdakis, B.; Cepic, Z.; Senk, N.; Doric, J.; Radovanovic, L. Use of scrap tires in cement production and their impact on nitrogen and sulfur oxides emissions. *Energy Sources* **2016**, *38*, 485–493. [CrossRef]
12. Tsiliyannis, C.A. Alternative fuels in cement manufacturing: Modeling for process optimization under direct and compound operation. *Fuel* **2012**, *99*, 20–39. [CrossRef]
13. Bingrong, L.; Wei, Y.; Peng, C.; Zhou, H.; Li, S.; Wang, S. Influence of Fe_2O_3 on the hydration kinetics of tricalcium aluminate. *J. Eur. Ceram. Soc.* **2023**, *43*, 6550–6561.
14. Chen, S.; Chen, G.; Cheng, J. Effect of additives on the hydration resistance of materials synthesized from the magnesia-calcia system. *J. Am. Ceram. Soc.* **2000**, *88*, 1810–1812. [CrossRef]
15. Tian, Y.; Pan, X.; Yu, H.; Tu, G. Formation mechanism and crystal simulation of Na_2O-doped calcium aluminate compounds. *Trans. Nonferrous Met. Soc. China* **2016**, *26*, 849–858. [CrossRef]
16. Boikova, A.I.; Domanskii, A.I.; Paramonova, V.A.; Stavitskaya, G.P.; Nikushchenko, V.M. The influence of sodium oxide on the structure and properties of tricalcium aluminate. *Cem. Concr. Res.* **1977**, *7*, 483–491. [CrossRef]
17. Mondal, P.; Jeffrey, J.W. Crystal structure of tricalcium aluminate, $Ca_3Al_2O_6$. *Acta Crystallogr.* **1975**, *31*, 689–697. [CrossRef]
18. Nguyen, M.; Sokolář, R. Formation and influence of magnesium-alumina spinel on the properties of refractory forsterite-spinel ceramics. *Mater. Technol.* **2020**, *54*, 135–141. [CrossRef]
19. Gobbo, L.; Sant'Agostino, L.; Garcez, L. C_3A polymorphs related to industrial clinker alkalies content. *Cem. Concr. Res.* **2004**, *34*, 657–664. [CrossRef]
20. Dietmar, S.; Wistuba, S. Crystal structure refinement and hydration behaviour of doped tricalcium aluminate. *Cem. Concr. Res.* **2006**, *36*, 2011–2020.
21. Nishi, F.; Takeuchi, Y. Aluminum oxide (Al_6O_{18}) rings of tetrahedra in the structure of sodium calcium aluminate ($Ca_{8.5}NaAl_6O_{18}$). *Acta Crystallografica* **1975**, *31*, 1169–1173. [CrossRef]
22. Götz-Neunhoeffer, F.; Neubauer, J. Crystal structure refinement of Na-substituted C_3A by Rietveld analysis and quantification in OPC. In Proceedings of the 10th International Congress on the Chemistry of Cement, Gothenburg, Sweden, 2–6 June 1997.
23. Varma, S.P.; Wall, C.D. A monoclinic tricalcium aluminate (C_3A) phase in a commercial Portland cement clinker. *Cem. Concr. Res.* **1981**, *11*, 567–574. [CrossRef]
24. Singh, V.K.; Ali, M.M.; Mandal, U.K. Formation Kinetics of Calcium Aluminates. *J. Am. Ceram. Soc.* **1990**, *73*, 872–876. [CrossRef]

25. Mohamed, B.M.; Sharp, J.H. Kinetics and mechanism of formation of tricalcium aluminate, $Ca_3Al_2O_6$. *Thermochim. Acta* **2002**, *388*, 105–114. [CrossRef]
26. López, F.A.; Martín, M.I.; Alguacil, F.J.; Sergio Ramírez, M.; González, J.R. Synthesis of Calcium Aluminates from Non-Saline Aluminum Dross. *Materials* **2019**, *12*, 1837. [CrossRef] [PubMed]
27. Pati, R.K.; Panda, A.B.; Pramanik, P. Preparation of Nanocrystalline Calcium Aluminate Powders. *J. Mater. Synth. Process.* **2002**, *10*, 157–161. [CrossRef]
28. Mandic, V.; Kurajica, S. The influence of solvents on sol–gel derived calcium aluminate. *Mater. Sci. Semicond. Process.* **2015**, *38*, 306–313. [CrossRef]
29. Ghoroi, C.; Suresh, A.K. Solid–solid reaction kinetics: Formation of tricalcium aluminate. *AIChE J.* **2007**, *53*, 502–513. [CrossRef]
30. Shin, G.Y.; Glasser, F.P. Interdependence of sodium and potassium substitution on tricalcium aluminate. *Cem. Concr. Res.* **1982**, *13*, 135–170. [CrossRef]
31. Ostrowski, C.; Żelazny, J. Solid Solutions of Calcium Aluminates C3A, C12A7 and CA with Sodium Oxide. *J. Therm. Anal. Calorim.* **2004**, *75*, 867–885. [CrossRef]
32. Scherrer, P. Bestimmung der Grosse und der Inneren Struktur von Kolloidteilchen Mittels Rontgenstrahlen. *Nachrichten Von Der Ges. Der Wiss. Math.-Phys. Kl.* **1918**, *2*, 98–100.
33. Warren, B.E. *X-ray Diffraction*; Addison-Wesley: Boston, MA, USA, 1969.
34. Ravaszová, S.; Dvořák, K. Development of Crystallinity of Triclinic Polymorph of Tricalcium Silicate. *Materials* **2020**, *13*, 3734. [CrossRef]

buildings

MDPI

Article

Early-Age Shrinkage Stress of Alkali-Activated Cement-Free Mortar Using Shrinkage Reducing Agent and Expansive Additive

Seok-ho Yoon [1], Sung-rok Oh [2], Ji-young Kim [3] and Sung Choi [1,*]

[1] Department of Civil Engineering, Kyungdong University, 27, Gyeongdongdaehak-ro, Yangju-si 11458, Gyeonggi-do, Republic of Korea; shyoon@kduniv.ac.kr
[2] Department of Civil Engineering, Semyung University, 65, Semyeong-ro, Jecheon-si 27136, Chungcheongbuk-do, Republic of Korea; cgdbs02@nate.com
[3] Smart Hi-Tech Group, Samsung E&A, 26, Sangil-ro 6 gil, Gangdong-gu, Seoul 05288, Republic of Korea; jy0421.kim@samsung.com
* Correspondence: csomy1113@kduniv.ac.kr

Abstract: Cement-free concrete has a superior physical performance, such as in its strength and durability, compared to OPC concrete; however, it has the disadvantage of large shrinkage. Large shrinkage can cause cracks due to shrinkage stress in the long term. In this study, a shrinkage reducing agent (SRA) was used to reduce the shrinkage of cement-free mortar; its content was increased from 0.0 to 1.5%. For an SRA content of 1.0%, a calcium sulfoaluminate (CSA) expansive additive (EA) (2.5, 5.0, and 7.5%) was added. To calculate the shrinkage stress of cement-free mortar using the SRA and EA, the compressive strength, elastic modulus, and total and autogenous shrinkage were measured. The unit shrinkage stress of cement-free mortar was obtained by multiplying the elastic modulus by the length change and accumulated to obtain the shrinkage stress acting on the mortar according to the age. The shrinkage stress of cement-free mortar showed different tendencies as the age increased. At early ages, the shrinkage rate of the mortar occupied a large proportion of the shrinkage stress. In the long term, the shrinkage stress was significantly affected by the elastic modulus. As a result, SRA was found to be effective in reducing the shrinkage stress by decreasing both the elastic modulus and shrinkage. However, EA increased the shrinkage stress over the long term due to an increase in the elastic modulus even though it compensated for early-ages shrinkage.

Keywords: cement-free mortar; CSA expansive additive; elastic modulus; shrinkage reducing agent; shrinkage stress

1. Introduction

Cement, the main material for developing concrete performance, emits a large amount of carbon dioxide (CO_2) during its production process. Thus, as part of efforts to realize carbon neutrality, various studies have been conducted to reduce CO_2 emissions in the cement industry [1–7]. Among them, the most actively researched topic is cement-free binder, which does not use cement at all. The materials typically used in the research on cement-free binder are blast-furnace slag (BFS) and fly ash (FA), which are industrial by-products. By simultaneously using these materials and an appropriate amount of alkali activator to induce the reaction, a similar performance to ordinary Portland cement (OPC)-based mortar and concrete can be achieved. In this way, cement-free binder using industrial by-products is an eco-friendly material that can replace existing cement, and its applicability has been verified through many studies [8–11].

Cement-free mortar has the advantages of high strength at early ages, excellent chloride resistance, chemical resistance, and durability [12]. However, cement-free mortar tends to undergo a large shrinkage process, which also has a problem of increasing the risk of cracking [13]. The shrinkage of cement-free mortar mostly occurs at early ages. Over the long term, the shrinkage increment decreases; however, it continuously occurs and increases the risk of cracking. To address these problems, studies have been conducted on the shrinkage characteristics of cement-free mortar. Cartwright et al. [14] prepared alkali-activated slag (AAS) mortar using various activators and measured autogenous and drying shrinkage. They found that AAS causes up to six times higher shrinkage than OPC. Collins et al. [15] reported that higher shrinkage occurs to cement-free mortar because it is subjected to higher capillary stress as its pore size is smaller than that of OPC. The cracks caused by the shrinkage of cement-free mortar are considered to be the most serious problem in applying the mortar to structures [13,16]. Cement-free binders have excellent physical properties and high durability; however, this applies when there is no crack in the matrix of the cement-free mortar and concrete. The application of cement-free binders to structures can result in shrinkage leading to cracks depending on the constraints and curing conditions. The generated shrinkage cracking may lead to uncertainty about the durability of cement-free mortar over the long term [17,18].

Various methods can be applied to control shrinkage cracking. Among these, shrinkage reducing agents (SRAs) and expansive additives (EAs) are commonly used as a material approach. In the study of Abolfathi et al. [19], the polypropylene (PP), steel fiber (SF), and hexylene glycol-based SRA were used for mitigating the shrinkage of alkali-activated slag and fly ash (AASF) mortars incorporating a recycled fine aggregate (RFA), and its characteristics were analyzed. As a result, although the initial compressive strength of AASF mortar incorporating SRA was reduced, a similar level of strength to the reference specimen was obtained at 28 days; in addition, the free and restrained shrinkage were mitigated compared to the specimens incorporating PP and SF. Al Makhadmeh and Soliman [20] established the mechanisms of SRA by conducting research on the behavior of SRA and its interaction with the pore solution and systems of AAS. The high shrinkage of AAS systems is due to the high ratio of mesopores and the formation of silica gel with a high tendency to shrink. According to the results, SRA delayed the hydration and extended the setting time by reducing the dissolution affinity of ions, and reduced shrinkage caused by C-A-S-H syneresis by lowering the degree of the polymerization of C-A-S-H, the main hydrate. Xu et al. [21] used a MgO-based expansion agent (MEA) for mitigating the shrinkage of alkali-activated blast furnace slag–copper slag (AAS-CS). The addition of MEA exerted the filling effect and shrinkage-compensating effect by promoting the formation of hydrotalcite (Ht), an expansive hydrate, and the lowest shrinkage rate and highest compressive strength appeared when 4.0 wt% of MEA was added. Choi et al. [22] used CSA EA to compensate for the shrinkage of cement-free mortar and reported that the shrinkage stress of the mortar was not significantly reduced because it increased the elastic modulus as it reduced the shrinkage of the mortar.

In summarizing the existing research results, it can be seen that SRAs do not affect the mineralogical composition of cement-free mortar and could reduce shrinkage by altering the pore structure and surface tension depression [23]; however, they can also reduce the compressive strength by affecting the hydration reaction. On the other hand, EAs not only the reduce shrinkage of cement-free mortar by compensating for it with the expansion effect of hydrates, but they also have the effect of increasing the compressive strength or elastic modulus. When material supplementation is used to reduce the shrinkage of cement-free mortar, the factors that affect the characteristics of cement-free mortar as well as shrinkage must be comprehensively investigated, and technologies must be applied that can accurately predict the stress caused by shrinkage and reduce shrinkage in a timely manner. In this study, an SRA and EA were used to reduce the shrinkage of cement-free mortar, and the shrinkage stress acting on the mortar was analyzed. To calculate the shrinkage stress of cement-free mortar with an SRA and EA, the total and autogenous shrinkages were measured for 150 days, and the elastic modulus was measured at early ages (1 and 2 days of age) and at larger ages (28 and 91 days). In addition, a model equation for the elastic modulus of cement-free mortar according to the SRA and EA contents was established by applying the experimental values of the elastic modulus by age to the ACI 209 model equation. The use of the shrinkage of cement-free mortar and the model equation makes it possible to obtain the shrinkage stress of the mortar per unit time (unit shrinkage stress), and the cumulative shrinkage stress acting on the mortar at any age can be obtained by accumulating the unit shrinkage stress. The shrinkage stress of cement-free mortar varied depending on the SRA and EA contents used for shrinkage reduction; this was analyzed at an early age (2 days) and a later age (150 days).

2. Materials and Methods

2.1. Materials and Mixture Proportions of Cement-Free Mortar

Table 1 shows the physical properties and chemical compositions of the binders used in cement-free mortar (ground granulated blast furnace slag (GGBFS) and FA), admixtures used for shrinkage reduction (SRA and EA), and alkali-activator (A-Act.). GGBFS used as a binder satisfies the type 3 criteria of KS F 2563 [24], and FA meets the type 2 criteria of KS F 5405 [25].

Table 1. Physical properties and chemical composition of materials used.

Type	Density (kg/m^3)	Blaine (m^2/kg)	CaO (%)	SiO_2 (%)	Al_2O_3 (%)	Fe_2O_3 (%)	SO_3 (%)	MgO (%)	K_2O (%)	Na_2O (%)	L.O.I. (%)
GGBFS	2908	468	43.4	34.6	14.3	0.6	5.0	5.1	0.5	0.2	−0.2
FA	2203	322	3.5	56.8	22.8	6.9	0.5	1.8	1.1	0.8	2.4
SRA	3160	-	1.4	29.0	0.2	0.1	-	0.1	-	-	69.2
EA	2862	375	36.3	30.1	24.5	1.6	5.1	1.3	0.6	0.3	0.3
A-Act.	1026	-	-	46.3	-	-	-	-	-	50.5	-

The GGBFS was produced from POSCO steelworks (Pohang, Republic of Korea) and its major chemical components were CaO (43.4%), SiO_2 (34.6%), and Al_2O_3 (14.3%). The basicity coefficient (K_b = (CaO + MgO)/(SiO_2 + Al_2O_3)) of the GGBFS was 0.99, which was close to the neutral value for ideal alkali activation (1.0). Its hydration modulus (HM = (CaO + MgO + Al_2O_3)/SiO_2) was 1.82, which is higher than 1.4, the value for an excellent hydration reaction. The FA was a product obtained by processing the industrial by-products generated from KOREA MIDLAND POWER Co. (Boryeong, Republic of Korea). Its main chemical components were SiO_2 (56.8%), Al_2O_3 (22.8%), and Fe_2O_3 (6.9%) with a M_2O (K_2O + Na_2O) content of 1.9%. The EA was produced in Japan and was based on calcium sulfoaluminate (CSA); it contained CaO (36.7%), SiO_2 (30.4%), and Al_2O_3 (24.7%). The density and fineness of EA were 2.90 g/cm^3 and 3100 cm^2/g, respectively. The SRA was a white powder and was composed of SiO_2 and organic components. The SRA used in this study was glycols-based. The density and ash content of SRA were 3.16 g/cm^3 and 40%, respectively. The A-Act was a white water-soluble powder, and its main chemical components were SiO_2 and Na_2O (96.8%). The alkali activator was in the form of a white powder with a molar ratio of 0.95. In addition, alkali activators are manufactured separately by adjusting the chemical components. The ratio between SiO_2 and Na_2O was 0.92. Figure 1 shows the particle size distribution curve of the fine aggregate. The fine aggregate used in the mortar was river sand, which had a density of 2.51 g/cm^3, a water absorption of 1.09%, and a fineness modulus of 2.82.

Table 2 summarizes the mix proportions of cement-free mortar. A two-component binder in which GGBFS and FA were mixed at a ratio of 7:3 was used, and the A-Act content was 24% per unit of water content. Since the GGBFS:FA was fixed to 7:3 in the previous study [22] which used EA only, for intercomparison with it, the mix ratio of binder was determined to be the same as the previous study [22]. The water-to-binder (W/B) ratio was 0.451 and the sand-to-binder (S/B) ratio was 1.2.

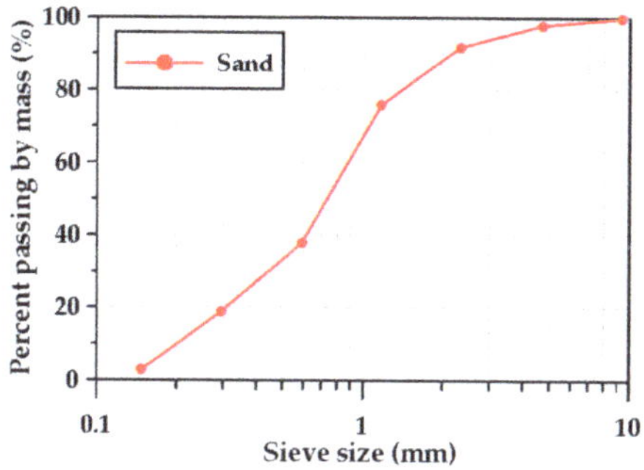

Figure 1. Particle size distribution curve of fine aggregates.

Table 2. Mix proportions of the cement-free mortar.

Type	Water (g)	Binder (g)	Activator (g)	Sand (g)	SRA (g)	EA (g)
S0.0	451	1000	108	1200	0	0
S0.5	451	1000	108	1200	2.26	0
S1.0	451	1000	108	1200	4.51	0
S1.5	451	1000	108	1200	6.77	0
S1E2.5	451	1000	108	1200	4.51	25
S1E5.0	451	1000	108	1200	4.51	50
S1E7.5	451	1000	108	1200	4.51	75

The SRA contents were 0.5, 1.0, and 1.5% per unit of water content. For the cement-free mortar that used both an SRA and EA, the SRA content was 1.0% and the EA content was 2.5, 5.0, and 7.5% compared to the unit binder content (since the dosage rate of EA had been 0, 2.5, 5.0, and 7.5% in the previous study [22], for association with it, the dosage rate of EA started from 2.5%). In total, seven types of cement-free mortar were prepared.

To prepare the cement-free mortar, the basic binder, A-Act, SRA, and EA were placed in a test container and dry-mixed for 30 s. The weighed water was added to the mixture and mixing was performed at 150 rpm for ten minutes to create a paste. Finally, the weighed fine aggregate was added, and mixing was performed at 300 rpm for 90 s.

2.2. Test Methods

2.2.1. Compressive Strength Test

Figure 2 shows the compressive strength test. The compressive strength test was conducted in accordance with ASTM C 39/M (2018) [26]. The specimens for this test were prepared using a cylindrical specimen mold with a diameter of 100 mm and height of 200 mm. The test was conducted at 1, 2, 28, and 91 days of age. The specimens were cured in a constant temperature and humidity chamber at a temperature of 23 ± 2 °C and a relative humidity of 90 ± 2% for 1 day, and then water-cured at same temperature until the ages of the compressive strength measurement. Before testing, the specimens were dried for 2 h to avoid the influence of pore water pressure. The test results of the three specimens were averaged and used as the compressive strength at each age.

Figure 2. The compressive strength test.

2.2.2. Modulus of Elasticity Test

Figure 3 shows the elastic modulus test. The elastic modulus test was conducted in accordance with ASTM C 469M-14 [27]. Specimens for this test were cylindrical with a diameter of 100 mm and a height of 200 mm. The test was conducted at 1, 2, 28, and 91 days of age. For the elastic modulus measurement, two strain gauges were installed on the side of each specimen. The test was conducted after checking the normal operation of the gauges by applying a load twice. After loading up to 40% of the ultimate load at a rate of 0.25 MPa/s, the elastic modulus was calculated through the regression of the deformation under the load using an interpolation method. The measurements of the three specimens were averaged and used as the elastic modulus. The specimens were cured in a constant temperature and humidity chamber at a temperature of 23 ± 2 °C and a relative humidity of 90 ± 2% for 1 day, and then water-cured at the same temperature until the ages of the modulus of elasticity measurement.

(**a**) (**b**) (**c**)

Figure 3. The elastic modulus test: (**a**) strain gauge; (**b**) test specimen; and (**c**) overall setup for measurement.

2.2.3. Shrinkage Test

Figure 4 shows the specimen for conducting the shrinkage test. To measure the length change rate of the mortar, 100 × 100 × 400 mm prismatic specimens were prepared. In the center of each specimen, an embedded gauge (PMFL-50-2LT, Tokyo Sokki Kenkyujo Co., Ltd., Tokyo, Japan) for the length change and a temperature sensor (Thermocouple t type, Tokyo Sokki Kenkyujo Co., Ltd., Tokyo, Japan) were installed. Before pouring the mortar, Teflon sheets and polystyrene boards were installed on the inner surface of the mold to minimize the friction between the mold and the constraint in the longitudinal direction. After pouring the mortar, polyester films were installed on the surface to prevent the evaporation and absorption of moisture on the surface of the specimen. The cement-free mortar specimens were demolded after one day of curing. For measuring the autogenous shrinkage, the specimens were sealed with polyester films to control drying. For measuring the total shrinkage, which is the sum of drying and autogenous shrinkage, the specimens were not sealed after demolding. The length change of the cement-free mortar was measured in a constant temperature and humidity chamber (temperature 20 °C and relative humidity 60%). The length change rate data were received every 30 min and the total measurement period was 150 days.

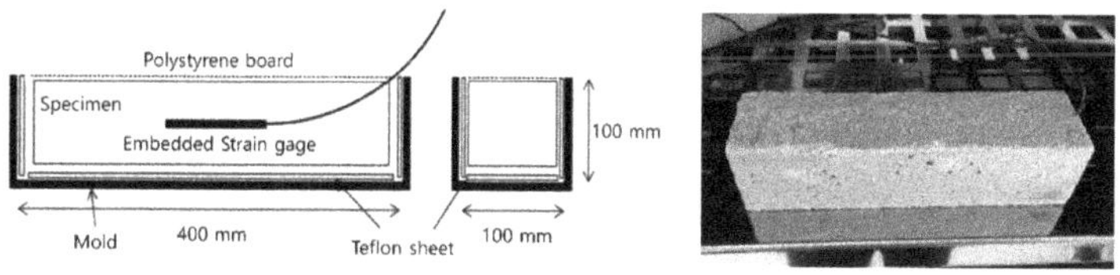

(a) (b)

Figure 4. The specimen for conducting shrinkage test: (**a**) schematic and (**b**) photograph.

3. Results

3.1. Compressive Strength

Table 3 shows the compressive strengths of the cement-free mortars that used the SRA and EA, which were used for shrinkage reduction. The binder of the cement-free mortar shows a fast reaction at the beginning due to the stimulation of the A-Act, and the strength rapidly increases at early ages [28,29].

Table 3. Compressive strength of cement-free mortar.

Age (Day)	S0.0 (MPa)	S0.5 (MPa)	S1.0 (MPa)	S1.5 (MPa)	S1E2.5 (MPa)	S1E5.0 (MPa)	S1E7.5 (MPa)
1	2.88	2.27	2.17	2.12	3.90	4.28	4.57
2	34.45	27.74	26.02	27.85	28.95	33.28	33.45
28	45.39	38.65	37.74	37.57	40.89	46.13	48.77
91	52.35	47.26	42.10	41.65	43.58	50.03	50.90

Meanwhile, the reaction rate of this hydration mechanism varies depending on the admixture used [30]. On day 1, the compressive strengths of the cement-free mortars that used the SRA were lower than that of S0.0, which did not use the SRA; the compressive strengths decreased as the content of the SRA increased. Bílek et al. [31] reported that an increase in the content of the SRA may intensify this strength reduction phenomenon because the SRA decreases the surface tension of mixing water to reduce the drying shrinkage, thereby delaying the ionic reaction of the A-Act dissolved in the mixing water and the alkali stimulation of the binder. In addition, Nguyen et al. [32] demonstrated that the reduction in the hydration reaction caused by an SRA may last for an extended period. In the experimental results of the present study, the cement-free mortar using the SRA showed lower compressive strengths at 28 and 91 days of age than S0.0.

The cement-free mortar using the SRA and EA exhibited higher compressive strengths than S0.0 on day 1, but its strengths were lower than that of S0.0 after day 2. The compressive strengths of the cement-free mortars using the EA increased on day 1 due to the expansion effect of the EA. The EA forms ettringite as a hydrate mainly at the early ages, which increases the compressive strength and has the effect of reducing shrinkage. Since this expansion effect of the EA was significant from days 1 to 2 and decreases afterwards, the strength reduction phenomenon of the EA occurred over the long term [33].

3.2. Modulus of Elasticity

Figure 5 shows the experimental values of the elastic modulus of the cement-free mortar using the SRA and EA and the prediction curve by the ACI 209 model equation [34].

The elastic modulus of the cement-free mortar using SRA showed no significant difference from that of S0.0 at day 1, but it was lower over the long term (after 28 days). This is because the cement-free mortar using the SRA has many pores [35,36]. On the other hand, the elastic modulus of the cement-free mortar using the SRA and EA was higher than that of S0.0 because the hydrates of the EA, which are expansive substances, filled the pores of the mortar and densified its inside, leading to the improved pore structure by the hydrates of the EA along with the generated hydrates of the mortar [37].

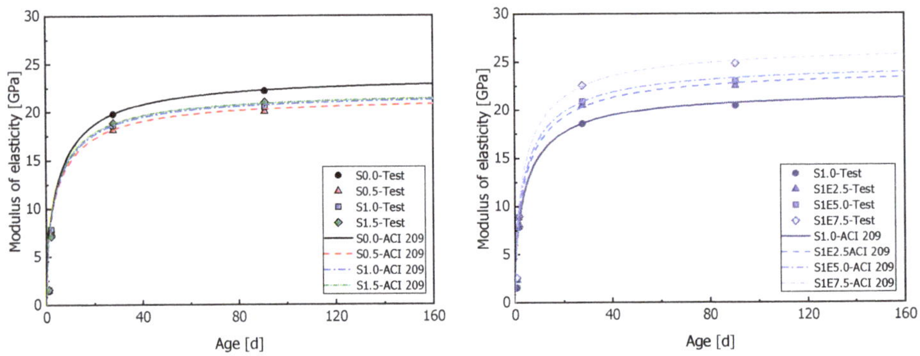

Figure 5. Comparison of the elastic modulus from the experimental values and the ACI 209 model equation: (**a**) usage of SRA and (**b**) usage of SRA and EA.

To analyze the change in the elastic modulus caused by the SRA and EA, a prediction curve was derived by applying the experimental values of the elastic modulus to the ACI 209 model equation which can predict the influence of the material used on the elastic modulus at early ages and at long ages, as follows:

$$E_{cmt} = E_{cm28}\sqrt{\frac{t}{a+bt}} \quad (1)$$

where E_{cmt} denotes the elastic modulus of cement-free mortar at age t, E_{cm28} denotes the elastic modulus at 28 days of age, and a and b denote the material constants.

Table 4 summarizes the material constants of the elastic modulus of the cement-free mortar using SRA and EA. The R^2 (coefficient of determination) between the experimental values and the ACI 209 model equation ranged from 0.915 to 0.949. Based on S0.0 with no admixture (SRA or EA), the change in the material constants of the elastic modulus caused by the use of the admixtures was analyzed. a is the constant for the maximum value of the elastic modulus. When the tendency of a was analyzed according to the contents of the admixtures used (SRA and EA), a decreased as the content of the SRA increased, but increased as the content of the EA increased when both the SRA and EA were mixed. This is because the long-term elastic modulus was high when the EA was used. b is the rate of change of the elastic modulus, which increases with age. b showed a tendency to increase as the content of the SRA increased. When both the SRA and EA were mixed, b showed a tendency to decrease as the content of the EA increased. The material constants a and b obtained by applying the elastic modulus of the cement-free mortar using the SRA and EA to the ACI 209 model equation can be used to predict the elastic modulus of the mortar at any age.

Table 4. Material constants and coefficient of determination of the elastic modulus of cement-free mortar.

Type	S0.0	S0.5	S1.0	S1.5	S1E2.5	S1E5.0	S1E7.5
a	8.496	8.122	8.020	7.831	8.456	8.735	9.016
b	0.697	0.710	0.714	0.720	0.698	0.688	0.678
R^2	0.915	0.918	0.919	0.920	0.945	0.949	0.945

3.3. Shrinkage

Figure 6 shows the total shrinkage (ε_{tsh}) and autogenous shrinkage (ε_{ash}) of cement-free mortar depending on the contents of the SRA and EA. Table 5 summarizes the shrinkage test results of the cement-free mortar using the SRA and EA at day 1, 2, and 150.

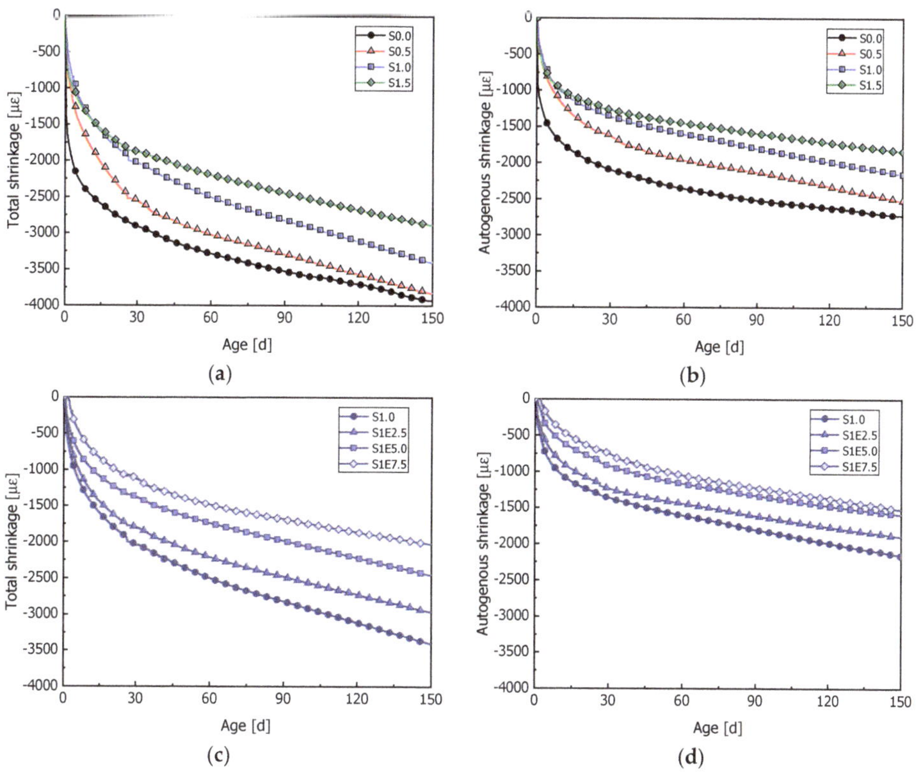

Figure 6. Shrinkage curves of cement-free mortar: (**a**) total shrinkage (SRA); (**b**) autogenous shrinkage (SRA); (**c**) total shrinkage (S1.0 + EA); and (**d**) autogenous shrinkage (S1.0 + EA).

For S0.0 without an SRA or EA, the total shrinkage was measured to be −3932 με on day 150. The total shrinkage was 41.6% (−1637 με) on day 1 and 47.1% (−1853 με) on day 2, indicating that significant shrinkage occurred on these days. At early ages, when significant shrinkage occurs, the elastic modulus increases, causing significant changes in the early-ages shrinkage stress that acts on cement-free mortar. The total shrinkage of the cement-free mortar using the SRA, however, was lower than that of S0.0 at early ages (0 to 2 days). The total shrinkage reduction rate increased as the content of the SRA increased. The total shrinkage of S1.5 with the highest content of the SRA was approximately 19.9% of that of S0.0 at day 1. The total shrinkage of S1.0 was also approximately 40.9% of that of

S0.0 at day 1. The total shrinkage for the case of mixing the EA with S1.0 was even lower than that of S1.0 at day 1. Particularly, S1E7.5, with the highest EA content, showed an expansion of 7 με. As the content of the SRA and EA increased, the total shrinkage at day 150 decreased, but the total shrinkage reduction rate was lower compared to day 1.

Table 5. ε_{ash} and $\varepsilon_{tsh} - \varepsilon_{ash}$ of cement-free mortar over 150 days.

Type	ε_{tsh} (με)			ε_{ash} (με)			$\varepsilon_{tsh} - \varepsilon_{ash}$ (με)		
	1 d	2 d	150 d	1 d	2 d	150 d	1 d	2 d	150 d
S0.0	−1637	−1853	−3932	−1096	−1237	−2728	−541	−616	−1204
S0.5	−822	−969	−3826	−437	−626	−2518	−385	−343	−1308
S1.0	−670	−838	−3401	−424	−554	−2154	−246	−284	−1247
S1.5	−326	−621	−2887	−264	−443	−1839	−62	−178	−1048
S1E2.5	−365	−485	−2960	−184	−335	−1896	−181	−150	−1064
S1E5.0	−109	−344	−2452	+11	−119	−1594	−120	−225	−858
S1E7.5	+7	−7	−2024	+29	+24	−1523	−22	−31	−501

Cement-free mortar also exhibited significant autogenous shrinkage at early ages. The autogenous shrinkage curve of S0.0 according to the age showed a tendency similar to that of the total shrinkage. The autogenous shrinkage at day 150 was 40.2% on day 1 and 45.3% on day 2. When the SRA was used, the autogenous shrinkage reduced. The autogenous shrinkage of S1.0 was approximately 38.7% of that of S0.0 on day 1 and the reduction rate was similar to that of the total shrinkage. When the EA was added to S1.0, the autogenous shrinkage reduced. Depending on the content of the EA, expansion was also observed. The cement-free mortar using the SRA and EA showed a reduction in autogenous shrinkage even at day 150, at which its reduction rate, compared to S1.0, showed a tendency similar to that of the total shrinkage reduction rate. In this instance, the difference between the autogenous and total shrinkage reduction rate was found to be less than 7.5%. The total shrinkage is the sum of the autogenous and drying shrinkage, and the SRA and EA, which are used to reduce the shrinkage of cement-free mortar, have the effect of reducing the total shrinkage [38–40].

To validate the shrinkage results obtained in this study, previous studies conducted under a similar topic were investigated. The total shrinkage at 150 days in this study was in the range of 2000–4000 με and this result was similar to [23], which had a similar range using the SRA up to 2%. In addition, the drying shrinkage at 56 days using the SRA and EA were approximately 1000 με and in the range of 500–1000 με, respectively, and these results were similar to [33]. It can be confirmed that although the exact results were slightly different due to the different types of the SRA and EA and mixture proportions, the range of results were similar between this study and previous studies. In a previous study, AAM mortar with the CSA EA showed shrinkage reduction at an early age due to the expansion characteristic of the CSA EA, but there was no further reduction after 3 days [22]. In this study, shrinkage reduction occurred for 150 days when the EA and SRA were used together, which we believe is due to the influence of the SRA.

3.4. Stress of Shrinkage (2 Days)

For cement-free mortar, significant shrinkage occurs at early ages. Such shrinkage may cause stress depending on the constraint conditions, and this stress may cause cracking at early ages when the strength of cement-free mortar is low [41]. The use of admixtures, such as the SRA and EA, can reduce the shrinkage of cement-free mortar. However, changes in the elastic modulus over time must be considered because the shrinkage stress is affected by both the shrinkage rate and the elastic modulus. Since both the shrinkage and elastic modulus of cement-free mortar change over time, the shrinkage stress ($f_{(sh,t+\Delta t)}$) is calculated by adding the stress caused by shrinkage per unit of time (Δf_{sh}) to the shrinkage stress acting on the cement-free mortar ($f_{(sh,t)}$) [22,42], as follows:

$$f_{(sh,t+\Delta t)} = f_{(sh,t)} + \Delta f_{sh} \quad (2)$$

$$\Delta f_{sh} = E \times \Delta \varepsilon_{sh} \quad (3)$$

Figure 7 shows the unit shrinkage stress and cumulative shrinkage stress of cement-free mortar until day 2. The unit shrinkage stress (Δf_{sh}) is the shrinkage stress calculated per unit of time and is obtained by multiplying the shrinkage rate, which varies every 30 min, by the elastic modulus. The cumulative shrinkage stress is the shrinkage stress acting on the cement-free mortar at time t and is obtained by adding the unit shrinkage stress to the shrinkage stress of the cement-free mortar.

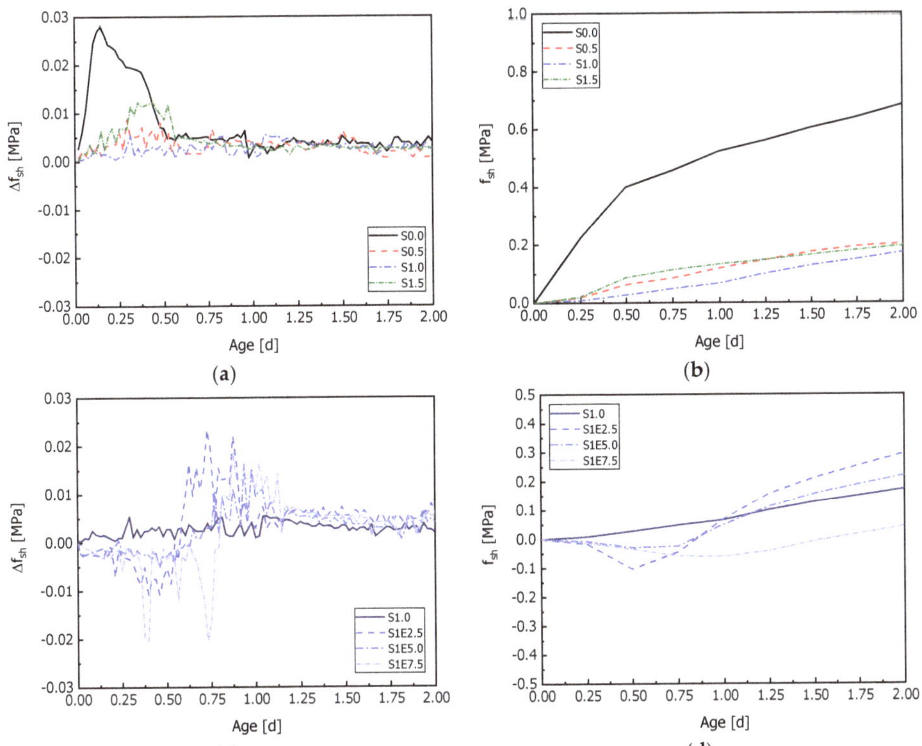

Figure 7. Shrinkage stress over 2 days caused by the shrinkage of cement-free mortar: (**a**) unit shrinkage stress (Δt: 30 min) according to usage of SRA; (**b**) cumulative shrinkage stress according to usage of SRA; (**c**) unit shrinkage stress (Δt: 30 min) according to usage of S1.0 and EA; and (**d**) cumulative shrinkage stress according to usage of S1.0 and EA.

The unit shrinkage stress of S0.0 increased from 0 to 0.146 days (3.5 h). The shrinkage stress reached a maximum of 0.028 MPa and then decreased until 0.5 days. Until 0.5 days, the unit shrinkage stress was high because the total shrinkage rapidly increased even though the elastic modulus was not yet large. However, after 0.5 days, the low unit shrinkage stress was maintained. The average unit shrinkage stress during the period from 0.5 to 2 days was approximately 0.004 MPa. Thus, the cumulative shrinkage stress of S0.0 increased until 0.5 days, and its increment decreased after 0.5 days.

The unit shrinkage stress of the cement-free mortar using the SRA was not large until 0.5 days. As the content of the SRA increased, the maximum unit shrinkage stress showed a tendency to decrease. This is because the shrinkage of the cement-free mortar using SRA was not high at the beginning and its elastic modulus was also low compared to that of S0.0. The unit shrinkage stress of the cement-free mortar using the SRA reached its peak at 0.25 to 0.31 days. The occurrence of the maximum unit shrinkage stress was delayed compared to S0.0 because the increase in the elastic modulus was delayed even though the cement-free mortar using the SRA also had the same tendency to exhibit significant shrinkage at the beginning.

The shrinkage of the cement-free mortar using the SRA and EA at early ages was further reduced and expansion occurred for a certain period. The unit shrinkage stress indicated that the expansion stress occurred for all specimens at the beginning and then shrinkage stress occurred. The duration of the expansion stress was shortest (11 h) for S1E2.5, which had the lowest content of the EA, and longest (16.5 h) for S1E7.5. The maximum value of the expansion stress was not proportional to the content of the EA because the time point at which expansion occurred and the time point at which the elastic modulus increased were different depending on the specimens. The cumulative shrinkage stress of S1E2.5 and S1E5.0 on day 2 exhibited larger values than S1.0 without the EA. This is because the elastic modulus was increased by the EA.

3.5. Stress of Shrinkage (150 Days)

The shrinkage stress of cement-free mortar is affected by the shrinkage rate and elastic modulus. As outlined in Section 3.4, the shrinkage significantly changes and has a large impact on shrinkage stress at early ages; however the influence of the elastic modulus increases as the age increases because the rate of the change of shrinkage decreases and the elastic modulus increases with time. Figure 8 shows the unit shrinkage stress and cumulative shrinkage stress of cement-free mortar until day 150.

The unit shrinkage stress results show that the unit shrinkage stress significantly varied until day 20. This is because the shrinkage measured by the data logger before day 20 had deviations of the unit time. After day 20, curves with a certain tendency were observed because the shrinkage of cement-free mortar did not significantly increase and thus the rate of the change of shrinkage per unit of time was small. In the unit shrinkage stress results over 150 days, S0.0 showed the highest shrinkage stress and the unit shrinkage stress of the cement-free mortar using the SRA decreased. The unit shrinkage stress shows a tendency to slightly increase after 20 days because the elastic modulus is large even though the shrinkage increment is small. At day 150, S0.0 showed the highest unit shrinkage stress (0.002397 MPa) and S1.5 showed the lowest unit shrinkage stress (0.001344 MPa).

The cumulative shrinkage stress results show that the shrinkage stress of S0.0 significantly increased at the beginning, and the cumulative shrinkage stresses rapidly increased until later ages. Meanwhile, the cumulative shrinkage stresses of the cement-free mortars using the SRA increased gently compared to that of S0.0. At day 150, S0.0 showed the highest cumulative shrinkage stress of 4.306 MPa. As the content of the SRA increased, the cumulative shrinkage stress decreased at day 150. Compared to S0.0, the cumulative shrinkage stress decreased by 40.7%, 42.4%, and 56.6% for S0.5, S1.0, and S1.5, respectively.

The unit shrinkage stress of the cement-free mortar made by adding the EA to S1.0 also significantly fluctuated until day 20, and the expansion section caused by the EA occurred at the beginning. The EA decreased the shrinkage of the cement-free mortar; however, a high unit shrinkage stress compared to S1.0 was observed over the long term due to the increase in the elastic modulus. When the unit shrinkage stress was compared in terms of the content of the EA, S1E2.5 exhibited the highest unit shrinkage stress (3.824 MPa) at day 150 followed by S1E5.0 (3.323 MPa) and S1E7.5 (3.157 MPa), but there was no significant difference in the unit shrinkage stress.

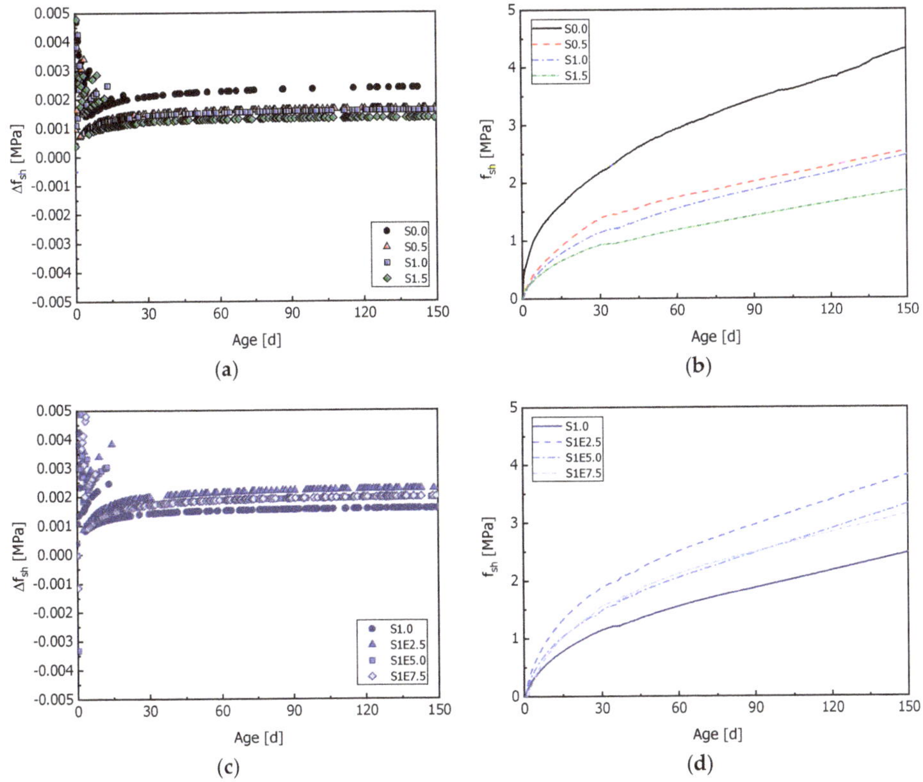

Figure 8. Shrinkage stress over 150 days caused by the shrinkage of cement-free mortar: (**a**) unit shrinkage stress (Δt: 30 min) according to usage of SRA; (**b**) cumulative shrinkage stress according to usage of SRA; (**c**) unit shrinkage stress (Δt: 30 min) according to usage of S1.0 and EA; and (**d**) cumulative shrinkage stress according to usage of S1.0 and EA.

4. Conclusions

An SRA and EA were used to reduce the shrinkage of cement-free mortar; the shrinkage stress was calculated using the compressive strength, elastic modulus, and shrinkage test results, and the shrinkage stress acting on the cement-free mortar at early ages and over the long term was analyzed. We draw the following conclusions from our results.

1. Using an SRA only, the compressive strengths decreased as the content of the SRA increased due to the low surface tension of the mixing water and a delay in the hydration reaction. When using both an SRA and EA, the initial compressive strengths were higher than those of S0.0 due to the expansion effect; however, this phenomenon was not continued over the long term.
2. At early ages (2 days), the unit shrinkage stress decreased, and the occurrence of the maximum unit shrinkage stress was delayed when using the SRA since it had the low elastic modulus compared to the S0.0 specimens. Meanwhile, the expansion occurred when using both the SRA and EA and the duration of the expansion increased as the EA content increased. In addition, the cumulative shrinkage stress was larger than S1.0 due to the increase in the elastic modulus by the EA.
3. At later ages (150 days), the influence of the elastic modulus increased the cumulative shrinkage stress. As the SRA content increased, the cumulative shrinkage stress decreased due to the low shrinkage rate and low elastic modulus. Meanwhile, the

cumulative shrinkage stress of the specimens incorporating the EA increased despite the reduced shrinkage rate due to the increase in the elastic modulus.

The use of the shrinkage stress monitoring technique based on the shrinkage and elastic modulus of mortar makes it possible to predict the long-term shrinkage stress of cement-free mortar and quantitatively evaluate the reduction in the shrinkage stress by expansive materials that are used for shrinkage reduction. In addition, the influence of shrinkage on the mechanical performance of structures can be evaluated through research on the shrinkage stress of structures that use cement-free binders according to constraint and curing conditions. In further research, for better understanding the microstructure change at early ages, the correlation between the shrinkage stress and microstructure of cement-free mortar should be investigated through X-ray diffraction (XRD), mercury intrusion porosimetry (MIP), gas pycnometry (GP), and scanning electron microscopy (SEM).

Author Contributions: Conceptualization, S.C., S.-h.Y. and S.-r.O.; investigation, S.-r.O. and J.-y.K.; resources, S.-h.Y.; writing—original draft, S.-h.Y. and S.-r.O.; writing—review and editing, S.-r.O. and J.-y.K.; supervision, S.-h.Y.; project administration, S.C.; funding acquisition, S.C. All authors have read and agreed to the published version of the manuscript.

Funding: This research received no external funding.

Data Availability Statement: Data are contained within the article.

Acknowledgments: This research was supported by a grant (22CTAP-C163834-02) from the Construction Technology Research Program funded by the Ministry of Land, Infrastructure, and Transport of the Korean government.

Conflicts of Interest: The authors declare no conflicts of interest.

References

1. Miller, S.A.; John, V.M.; Pacca, S.A.; Horvath, A. Carbon dioxide reduction potential in the global cement industry by 2050. *Cem. Concr. Res.* **2018**, *114*, 115–124. [CrossRef]
2. Barcelo, L.; Kline, J.; Walenta, G.; Gartner, E. Cement and carbon emissions. *Mater. Struct.* **2014**, *47*, 1055–1065. [CrossRef]
3. Hasanbeigi, A.; Price, L.; Lin, E. Emerging energy-efficiency and CO_2 emission-reduction technologies for cement and concrete production: A technical review. *Renew. Sustain. Energy Rev.* **2012**, *16*, 6220–6238. [CrossRef]
4. Das, S.; Saha, P.; Jena, S.P.; Panda, P. Geopolymer concrete: Sustainable green concrete for reduced greenhouse gas emission–A review. *Mater. Today Proc.* **2022**, *60*, 62–71. [CrossRef]
5. Yang, K.H.; Jung, Y.B.; Cho, M.S.; Tae, S.H. Effect of supplementary cementitious materials on reduction of CO_2 emissions from concrete. *J. Clean. Prod.* **2015**, *103*, 774–783. [CrossRef]
6. Wang, X.Y.; Lee, H.S. Effect of global warming on the proportional design of low CO_2 slag-blended concrete. *Constr. Build. Mater.* **2019**, *225*, 1140–1151. [CrossRef]
7. Elchalakani, M.; Basarir, H.; Karrech, A. Green concrete with high-volume fly ash and slag with recycled aggregate and recycled water to build future sustainable cities. *J. Mater. Civ. Eng.* **2017**, *29*, 04016219. [CrossRef]
8. Aliabdo, A.A.; Abd Elmoaty, M.; Aboshama, A.Y. Utilization of waste glass powder in the production of cement and concrete. *Constr. Build. Mater.* **2016**, *124*, 866–877. [CrossRef]
9. Choi, S.; Lee, K.M. Influence of Na_2O content and Ms (SiO_2/Na_2O) of alkaline activator on workability and setting of alkali-activated slag paste. *Materials* **2019**, *12*, 2072. [CrossRef]
10. Shubbar, A.A.; Sadique, M.; Shanbara, H.K.; Hashim, K. The development of a new low carbon binder for construction as an alternative to cement. In *Advances in Sustainable Construction Materials and Geotechnical Engineering*; Springer: Singapore, 2020.
11. Grist, E.R.; Paine, K.A.; Heath, A.; Norman, J.; Pinder, H. The environmental credentials of hydraulic lime-pozzolan concretes. *J. Clean. Prod.* **2015**, *93*, 26–37. [CrossRef]
12. Zhang, J.; Shi, C.; Zhang, Z.; Ou, Z. Durability of alkali-activated materials in aggressive environments: A review on recent studies. *Constr. Build. Mater.* **2017**, *152*, 598–613. [CrossRef]
13. Collins, F.; Sanjayan, J.G. Cracking tendency of alkali-activated slag concrete subjected to restrained shrinkage. *Cem. Concr. Res.* **2000**, *30*, 791–798. [CrossRef]
14. Cartwright, C.; Rajabipour, F.; Radlińska, A. Shrinkage characteristics of alkali-activated slag cements. *J. Mater. Civ. Eng.* **2014**, *27*, B4014007. [CrossRef]
15. Collins, F.G.; Sanjayan, J.G. Workability and mechanical properties of alkali activated slag concrete. *Cem. Concr. Res.* **1999**, *29*, 455–458. [CrossRef]

16. Gu, Y.; Fang, Y. Shrinkage, Cracking, Shrinkage-Reducing and Toughening of Alkali-Activated Slag Cement—A Short Review. *J. Chin. Ceram. Soc.* **2012**, *40*, 76–84.
17. Bernal, S.A.; Provis, J.L. Durability of alkali-activated materials: Progress and perspectives. *J. Am. Ceram. Soc.* **2014**, *97*, 997–1008. [CrossRef]
18. Law, D.W.; Adam, A.A.; Molyneaux, T.K.; Patnaikuni, I. Durability assessment of alkali activated slag (AAS) concrete. *Mater. Struct.* **2012**, *45*, 1425–1437. [CrossRef]
19. Abolfathi, M.; Omur, T.; Kabay, N. Effect of microfibers or SRA on the shrinkage and mechanical properties of alkali activated slag/fly ash-based mortars incorporating recycled fine aggregate. *Constr. Build. Mater.* **2023**, *373*, 130883. [CrossRef]
20. Al Makhadmeh, W.; Soliman, A. On the mechanisms of shrinkage reducing admixture in alkali activated slag binders. *J. Build. Eng.* **2022**, *56*, 104812. [CrossRef]
21. Xu, R.; Kong, F.; Yang, R.; Wang, H.; Hong, T. Drying shrinkage mitigation of alkali-activated blast furnace slag-copper slag by polyether-based shrinkage reducing admixture and MgO-based expansion agent. *Constr. Build. Mater.* **2024**, *416*, 135172. [CrossRef]
22. Choi, S.; Ryu, G.S.; Koh, K.T.; An, G.H.; Kim, H.Y. Experimental study on the shrinkage behavior and mechanical properties of AAM mortar mixed with CSA expansive additive. *Materials* **2019**, *12*, 3312. [CrossRef] [PubMed]
23. Palacios, M.; Puertas, F. Effect of shrinkage-reducing admixtures on the properties of alkali-activated slag mortars and pastes. *Cem. Concr. Res.* **2007**, *37*, 691–702. [CrossRef]
24. *KS F 2563*; Ground Granulated Blast-Furnace Slag for Use in Concrete. Korean Standards: Seoul, Republic of Korea, 2019.
25. *KS L5405*; Fly Ash. Korean Standards: Seoul, Republic of Korea, 2018.
26. *ASTM C39/C39M-18*; Standard Test Method for Compressive Strength of Cylindrical Concrete Specimens. ASTM International: West Conshohocken, PA, USA, 2016.
27. *ASTM C469M-14*; Standard Test Method for Static Modulus of Elasticity and Poisson's Ratio of Concrete in Compression. ASTM International: West Conshohocken, PA, USA, 2014.
28. Ravikumar, D.; Neithalath, N. Reaction kinetics in sodium silicate powder and liquid activated slag binders evaluated using isothermal calorimetry. *Thermochim. Acta* **2012**, *546*, 32–43. [CrossRef]
29. Chang, J.J. A study on the setting characteristics of sodium silicate-activated slag pastes. *Cem. Concr. Res.* **2003**, *33*, 1005–1011. [CrossRef]
30. Shi, C.; Roy, D.; Krivenko, P. *Alkali-Activated Cements and Concretes*; CRC Press: Boca Raton, FL, USA, 2003.
31. Bílek, V.; Kalina, L.; Novotný, R.; Tkacz, J.; Pařízek, L. Some issues of shrinkage-reducing admixtures application in alkali-activated slag systems. *Materials* **2016**, *9*, 462. [CrossRef]
32. Nguyen, C.V.; Mangat, P.S.; Jones, G. Effect of shrinkage reducing admixture on the strength and shrinkage of alkali activated cementitious mortar. In *IOP Conference Series: Materials Science and Engineering*; IOP Publishing: Bristol, UK, 2018; Volume 371, p. 012022.
33. Jia, Z.; Yang, Y.; Yang, L.; Zhang, Y.; Sun, Z. Hydration products, internal relative humidity and drying shrinkage of alkali activated slag mortar with expansion agents. *Constr. Build. Mater.* **2018**, *158*, 198–207. [CrossRef]
34. Vakhshouri, B.; Nejadi, S. Empirical models and design codes in prediction of modulus of elasticity of concrete. *Front. Struct. Civ. Eng.* **2019**, *13*, 38–48. [CrossRef]
35. Zhan, P.M.; He, Z.H. Application of shrinkage reducing admixture in concrete: A review. *Constr. Build. Mater.* **2019**, *201*, 676–690. [CrossRef]
36. Das, B.B.; Kondraivendhan, B. Implication of pore size distribution parameters on compressive strength, permeability and hydraulic diffusivity of concrete. *Constr. Build. Mater.* **2012**, *28*, 382–386. [CrossRef]
37. Speziale, S.; Jiang, F.; Mao, Z.; Monteiro, P.J.; Wenk, H.R.; Duffy, T.S.; Schilling, F.R. Single-crystal elastic constants of natural ettringite. *Cem. Concr. Res.* **2008**, *38*, 885–889. [CrossRef]
38. Coppola, L.; Coffetti, D.; Crotti, E.; Candamano, S.; Crea, F.; Gazzaniga, G.; Pastore, T. The combined use of admixtures for shrinkage reduction in one-part alkali activated slag-based mortars and pastes. *Constr. Build. Mater.* **2020**, *248*, 118682. [CrossRef]
39. Hu, X.; Shi, C.; Zhang, Z.; Hu, Z. Autogenous and drying shrinkage of alkali-activated slag mortars. *J. Am. Ceram. Soc.* **2019**, *102*, 4963–4975. [CrossRef]
40. Maltese, C.; Pistolesi, C.; Lolli, A.; Bravo, A.; Cerulli, T.; Salvioni, D. Combined effect of expansive and shrinkage reducing admixtures to obtain stable and durable mortars. *Cem. Concr. Res.* **2005**, *35*, 2244–2251. [CrossRef]
41. He, Z.; Li, Z.J.; Chen, M.Z.; Liang, W.Q. Properties of shrinkage-reducing admixture-modified pastes and mortar. *Mater. Struct.* **2006**, *39*, 445–453. [CrossRef]
42. Semianiuk, V.; Tur, V.; Herrador, M.F. Early age strains and self-stresses of expansive concrete members under uniaxial restraint conditions. *Constr. Build. Mater.* **2017**, *131*, 39–49. [CrossRef]

buildings

MDPI

Article

Three-Dimensional Printable Concrete by an Ultra-Thin Nozzle and Fully Sealed Extrusion

Jing Shen [1], Yujia Li [2], Xiaoman Zhang [1], Yangbo Li [2], Chaohui Huang [3] and Wei Luo [3,*]

1 School of Transportation Science and Engineering, Beihang University, Beijing 100191, China; chenjing0813@buaa.edu.cn (J.S.); zy2213209@buaa.edu.cn (X.Z.)
2 College of Hydraulic and Environmental Engineering, China Three Gorges University, Yichang 443002, China; lyj121341046@163.com (Y.L.); liyangbo@ctgu.edu.cn (Y.L.)
3 Haining Yancang River Embankment Management Office, Haining 314400, China; 13867363305@163.com
* Correspondence: luo810317@sohu.com; Tel.: +86-13758373234

Abstract: Due to the molding-free property and dry shrinkage of extrusion-based three-dimensional printable concrete (3DPC), the precision issues of 3DPC have not been solved effectively. One of the viable solutions for 3DPC precision improvement is to print using ultra-thin filaments. The challenges of ultra-thin-filament printing are extrudability, flowability, and fast solidification. To overcome these challenges and enhance precision, a customized 3D concrete printer with an ultra-thin diameter nozzle (6 mm) and fully sealed extrusion system was developed, and the mix design of ultra-thin-filament 3DPC (UTF-3DPC) was studied, including ingredients such as fly ash (FA), silica fume (SF), ordinary Portland cement (OPC), sodium dodecyl sulfate and cellulose (SDSC), water reducer, water, and sand. The function of UTF-3DPCs flowability and fast solidification with the proportion of water and SDSC was explored to obtain the optimal mix design. The standard compressive and flexural strengths of UTF-3DPC specimens were compared with the mold-cast vibrated and the mold-cast non-vibrated concrete. Their meso-scale and micro-scale structures were analyzed to expose the strength mechanism, according to the scanning electron microscope (SEM) images. A suitable mix design of UTF-3DPC was obtained and UTF-3DPC strength reached 80% of standard mold-cast concrete. The findings reported here provide a pathway to improve the precision of 3DPC and extend the application of 3D printing technology in engineering.

Keywords: 3D printable concrete; ultra-thin filaments; flowability; fast solidification; strength; microstructure

1. Introduction

3D printed concrete (3DPC) has been increasingly applied in the construction of under-5-story houses [1,2], post-tensioned segmental bridges [3,4], facilities in gardens or parks [5,6], complex templates [7], etc. worldwide. Engineering applications of 3DPC benefit us with intelligent fabrication [8], being molding-free, less labor and material consumption [9], being eco-friendly [10,11], etc. However, some barriers still hinder extensive applications in civil engineering, such as lack of reinforced concrete printing, full-graded concrete printing, and quality standards/specifications. Especially, we lack a series of standards or specifications to evaluate the precision, strength, durability, and anti-crack ability of 3DPC products including indoor, on-site experimental, and real-scale structures.

Among those issues, the precision of extrusion-based 3DPC is a non-negligible factor for structural security and integrity, which is mainly influenced by the configuration of the printer, such as filament, nozzle size, XYZ motion systems accuracy, and too significant dry shrinkage. Generally, filament or nozzle size is the primary factor for precision and surficial flatness. As shown in Figure 1, the thicker filament causes inevitable voids and surficial unevenness. The thinner the filament or nozzle is, the higher precision is and the better surficial flatness. Usually, the thinner the nozzle and filament are, the worse the extrudability of the concrete slurry has.

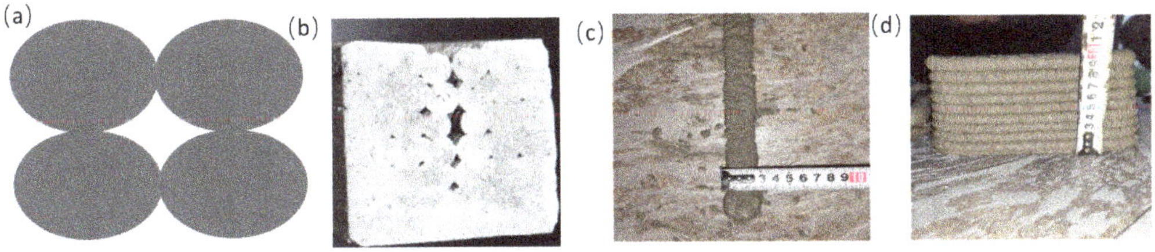

Figure 1. Void may occur during path planning (**a**), and void appears during actual printing (**b**). A 15 mm size is used for printing; the extrusion width changes to 25 mm (**c**). The designed layer height is 105 mm, and the actual layer height is 100 mm (**d**).

As shown in Table 1, current research based on printers with larger than 10 mm nozzles has been mostly reported. One exceptional example is cement-based materials printing by a printer with a 1.63 mm nozzle [12], where the aggregates are absent. There are few reports that the concrete printer with smaller than 10 mm nozzle succeeded in the research community, because the concrete slurry is hard to be extruded out continuously and fast solidify in a short time by a printer with thinner than 10 mm diameter nozzle, even though these admixtures, such as nano attapulgite clay [13], sodium carboxymethyl cellulose water-soluble polymer powder [14], hydrous magnesium aluminosilicate nanoclay [10], kaolinite clay [15], nano clay [16], cellulosic materials [17], polyacrylamide aqueous solution [18], are doped in to increase viscosity and reduce the yield stress. However, too small yield stress will reduce the buildability of 3DPC. Yield stress and viscosity are used to define the flowability of 3DPC. Besides the flowability, the fast solidification is also important for 3DPC. After extrusion, the deposited slurry should have sufficient strength to support the continuously constructed layer without excessive deformation or damage.

Table 1. Print nozzle diameter and research contents used by other researchers.

Author	The Size of Nozzle	Research Results
Bai, G., et al. [11]	30 mm	Explore the feasibility of eco-friendly concrete
Li Wang, et al. [19]	38 mm	Explore high belite sulfoaluminate and medium-heat Portland cements for large-scale 3D printing
Ma, et al. [20]	10 mm	Interlayer reinforcement in the process of printing.
L. Wang, et al. [21]	12 mm	Realize the real-time quantification of fresh and hardened mechanical property for 3D printing material
Jay G. Sanjayan, et al. [22]	25 × 10 mm rectangle	Explore the effect of surface moisture on inter-layer strength of 3D printed concrete
T.T. Le, et al. [23]	9 mm	Research on the hardened properties of high-performance printing concrete
A. Perrot, et al. [24]	35 mm circular nozzle and 21 × 42 mm rectangular nozzle	Assess the possibility of adapting the 3D printing technique to earth-based material.
Suvash Chandra Paul, et al. [25]	10 × 20 mm rectangular nozzle and 8 mm diameter nozzle	Explore the fresh and hardened properties of 3D printable cementitious materials for building and construction
Jian Hui Lim, et al. [26]	13 × 30 mm rectangular nozzle	Research on flexural characteristics of 3D printed geopolymer composites with in-process steel cable reinforcement

Table 1. *Cont.*

Author	The Size of Nozzle	Research Results
Biranchi Panda, et al. [27]	20 × 20 mm square nozzle and 15 × 30 mm rectangular nozzles.	Measurement of tensile bond strength of 3D printed geopolymer mortar
G. Ma, et al. [28]	8–30 mm nozzle	Printable properties of cementitious material containing copper tailings
Arun R. Arunothayanet, et al. [10]	30 mm nozzle	Digital fabrication of eco-friendly ultra-high performance fiber-reinforced concrete
Shin Hau Bong, et al. [5]	30 mm nozzle	Ambient temperature cured 'just-add-water' geopolymer for 3D concrete printing applications
Panda, B., et al. [13]	30 × 15 mm	Improving the 3D printability of high volume fly ash mixtures via the use of nano attapulgite clay
Sadegh Ghourchian, et al. [29]	150 × 50 mm	Modelling the development of capillary pressure in freshly 3D-printed concrete elements
Chao Liu, et al. [30]	20 mm nozzle	Influence of hydroxypropyl methylcellulose and silica fume on stability, rheological properties, and printability of 3D printing foam concrete
Nan Zhang, et al. [31]	240 mm nozzle	Short-duration near-nozzle mixing for 3D concrete printing
Bing Lu, et al. [32]	25.4 mm nozzle	Effect of printing parameters on material distribution in spray-based 3D concrete printing
Brahim Mazhoud, et al. [33]	21 × 40 mm^2 rectangular nozzle	Underwater 3D printing of cement-based mortar
Lowke, D., et al. [34]	8 mm diameter nozzle	Injection 3D concrete printing in a carrier liquid

To enhance the precision of 3DPC, a customized concrete 3D printer with 6 mm diameter nozzle and fully sealed extrusion has been developed. To successfully fabricate the ultra-thin filament 3DPC (UTF-3DPC), fly ash (FA), silica fume (SF), ordinary Portland cement (OPC), sodium dodecyl sulfate and cellulose (SDSC) at an optimal ratio are introduced to mix fine aggregates and water. The flowability and fast solidification of UTF-3DPC are measured to optimize the yield stress and viscosity. The standard samples are printed by our customized printer. The compressive and flexural strengths of UTF-3DPC have been tested and compared with ordinary standard mold-cast-vibrating and mold-cast-non-vibrating specimens. To expose the printability and strength mechanism of 3DPC, SEM imaging has been conducted. In previous research [35], the impact of different printing paths on the performance of 3DPC has been studied.

2. Material, Mix Design, Fabrication, and Testing Methods

2.1. Customized Concrete 3D Printer

To attain the robust fabrication of UTF-3DPC, we have developed our customized concrete 3D printer and fabricated specimens. Our customized concrete 3D printer is characterized by an ultra-thin nozzle with the diameter of 6 mm. Its benefits include higher accuracy and better surface flatness. It consists of an entire framework, XYZ motion system, extrusion system, slurry conveying system, mainboard, and embedded operating system as shown in Figure 2a,c. The entire framework is combined by base tray and gantry crane with the size of 500 mm long, 500 mm wide, and 700 mm high. The extruder in Figure 2b is mounted on the beam in the gantry crane, which controls X and Z axis movement, while the base tray moves along the Y axis. All X, Y, and Z axes movements in the XYZ motion system are driven by stepper motors and screw rods. The extruding system is featured by the fully sealed extrusion, including the extruder, plastic tube, and material tank. The concrete slurry sealed in the material tank is propelled by the conveying system into plastic tube and extruded out through the nozzle on the extruder. The slurry conveying system includes air compressor and the piston in the material tank. The mainboard and embedded operating system are used to control nozzle or trays moving path as given route. The maximum printing size of the specimens is 200 mm × 200 mm × 180 mm. The cylindrical material tank with an inner diameter of 50 mm is placed vertically next to the printer to ensure smooth propelling. The plastic tube with an inner diameter of 8 mm is used to deliver slurry into the extruder. An air compressor provides the high-pressure air to propel the piston at a stable velocity.

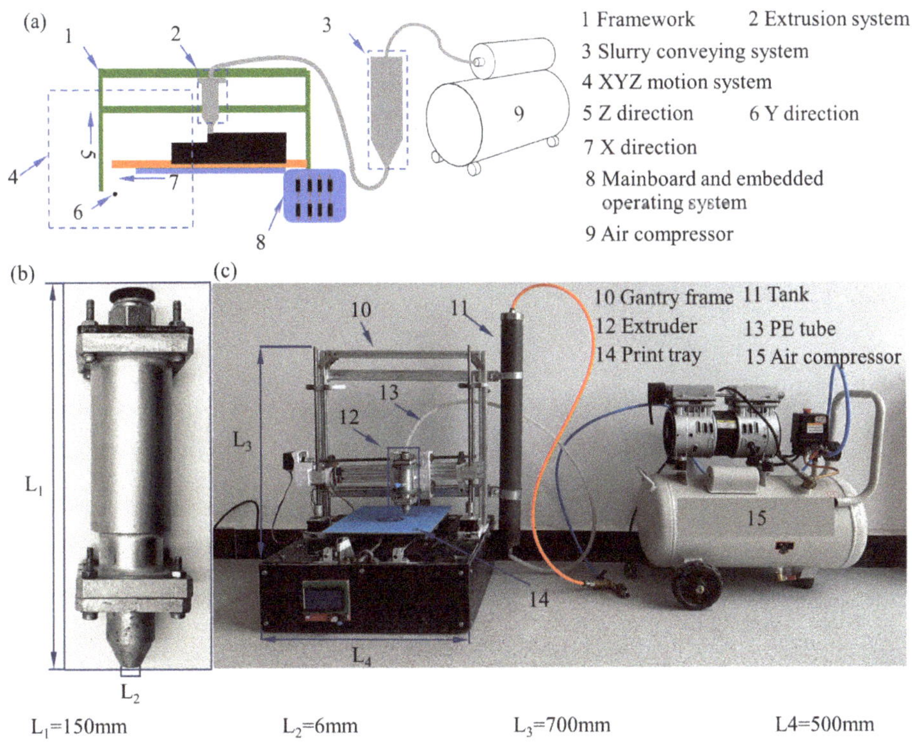

Figure 2. (**a**) The model diagram of the customized ultra-thin nozzle printer. It consists of entire framework, XYZ motion system, extrusion system, slurry conveying system, mainboard, and embedded operating system. (**b**) The extruder with 150 mm height and a 6 mm ultra-thin nozzle. (**c**) The real image of customized ultra-thin nozzle concrete 3D printer.

2.2. Materials and Mix Proportion

The optimal admixtures and mix design have been explored to ensure extruding the slurry out. To extrude the slurry smoothly, ingredients such as fly ash (FA), silica fume (SF), ordinary Portland cement (OPC), sodium dodecyl sulfate, and cellulose (SDSC) at a certain ratio are introduced to mix natural sand, water reducer, and water, and to fabricate the concrete specimens. The SEM images of the main ingredients are shown in Figure 3. The magnitude of 1000 times witnesses 2–10-micron spherical particles of FA in Figure 3a. SF behaves as 200–500-nanometer spherical particles in Figure 3b, while OPC as angular particles with the size of 1–50 micron in Figure 3c, and SDSC as angular particles with the size of 10–50 microns as well as fiber of 30 microns long in Figure 3d. The natural sand ranges from 0.08 mm to 2 mm.

Figure 3. SEM images of such main ingredients as (**a**) FA, (**b**) SF, (**c**) OPC, and (**d**) SDSC.

As a supplementary material for OPC, FA hydrates together with OPC, and adheres other ingredients and aggregates. Since the nozzle diameter of the customized concrete 3D printer is smaller than the nozzle diameters mentioned in most papers, more attention needs to be paid to the study of the extrudability and flowability of the concrete slurry. To be extruded smoothly, SF and SDSC are introduced to reduce the shear force between particles and the friction of the slurry and the nozzle or pipe. Also, SDSC is employed as a water-retaining material to lock the water inside the slurry under high pressure and avoid bleeding water out of the slurry. SDSC carrying water also feeds cementitious material with water to extend the hydration process. As for the admixture, polycarboxylic acid high-performance water-reducing agent is used as water reducer to control the workability of the slurry. After previous research [35] and experiments, the ratio of fine aggregates and cementitious materials is set to approximately 0.5. To explore optimal water or SDSC, six mix designs include W550SDSC14, W600SDSC14, W650SDSC14, W600SDSC7, W600SDSC14, and W600SDSC21 as shown in Table 2.

The chemical composition of the materials used is identical to that used in previous research [35].

Table 2. Mix ratios design of flowability test (kg/m^3).

Mix Ratios Design	Water	OPC	FA	SF	Sand	Water Reducer	SDSC
W550SDSC14	550						14
W600SDSC14	600	800	400	50	800	0.12	14
W650SDSC14	650						14
W600SDSC7	600						7
W600SDSC14	600	800	400	50	800	0.12	14
W600SDSC21	600						21

2.3. Specimen Fabrication

According to mix designs, all ingredients are put together, stirred evenly to obtain the slurry. Then the slurry is infused into the cylindrical tank. Before being sealed in the tank, air in the pipe and tank should be discharged out to ensure continuous extrusion. After 3D models of specimens are sliced into layers and filaments as shown in Figure 4a, printing paths from bottom to top are generated, which guide the nozzle to move and simultaneously extrude slurry out at a certain velocity under high-pressure air produced by air compressor. As shown in Figure 4b, 6 mm-diameter filaments are stacked into a layer. After stacking the current layer, the nozzle will move upward 5 mm to print a new layer until the specimen in Figure 4c is printed out. To print concrete specimens smoothly, 6 mm diameter filament or nozzle and 30 mm/s of nozzle moving velocity have been finally adopted. Every single printing layer is 5 mm high. The infill density is set as 100%. The air pressure of delivering the slurry is set as 0.4 MPa.

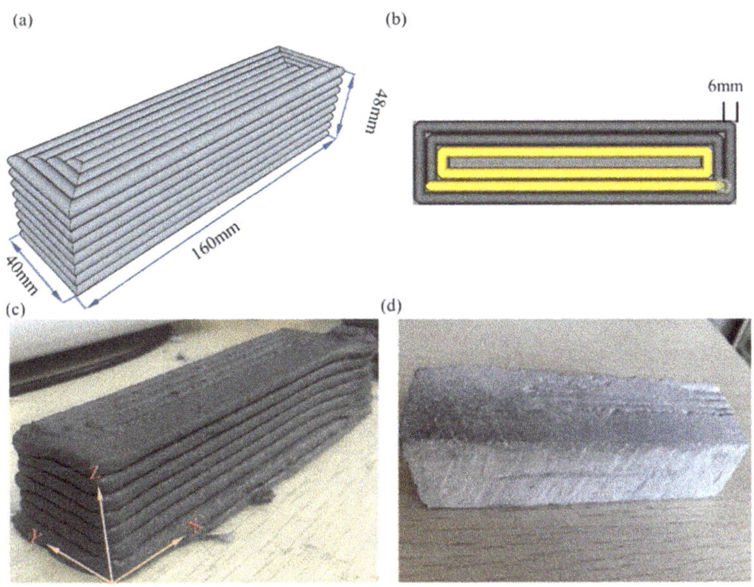

Figure 4. The sliced 3D model (**a**), filaments in a layer (**b**), printed specimen (**c**), and polished specimen of UTF-3DPC (**d**).

After curing for 28 days in a standard thermal and humid environment, specimens are polished into standard concrete specimens (40 mm × 40 mm × 160 mm), as shown in Figure 4d.

2.4. Flowability and Fast Solidification Testing

2.4.1. Flowability

According to multiple standards, GB/T 2419-2005 [36], several methods are introduced to test the flowability of concrete slurry, including the consistometer [5,11] and flow table test [11,28]. The consistometer is used here because the penetrating cone bears friction and resistance of the concrete slurry, which is similar to 3DPC.

During the experiment, the mixed mortar was placed in the consistometer, and the depth of the cone barrel of the consistency instrument was 150 mm. The mortar inside the cone barrel was punctured and the depth that the cone punched can be measured in the consistency. By this method, the performance of the mortar with different mix ratios is tested first, and the material performance is evaluated and printed after the relevant data is obtained.

Flowability plays a significant role in 3DPC. During the 3D printing process, concrete requires a lower yield stress and an appropriate plastic viscosity. The yield stress is computed from the depth of penetration h of the standardized cone used in this study (tip angle θ = 30°, mass of m = 300 g) as Equation (1) [33]:

$$\tau_0 = \frac{m \times g \times cos^2\theta}{\pi h^2 \times tan\theta} \quad (1)$$

where g is the gravity acceleration.

The dosage of water and the amount of SDSC are the main factors which affect the flowability. Because the printer nozzle is ultra-thin (6 mm), even a small change in the amount of water will have a great impact on the flowability of the slurry. As shown in Table 1, 5 mix designs are adopted according to the controlled variable method. To find out optimal water or SDSC proportion, mix designs W550SDSC14, W600SDSC14, and W650SDSC14 take water as variable floating ±50 g, and mix designs W600SDSC7, W600SDSC14, W600SDSC21 take SDSC as variable floating ±7 g. Three specimens are made in the same mix design.

2.4.2. Fast Solidification

According to ASTM GB/T 2419-2005, the velocity of solidification is depicted by the time of initial solidification, which is tested by the penetrometer. For successful printing, the materials used in 3DPC should be continuously extruded through the nozzle without any blockage or segregation. After extrusion, the deposited material should have sufficient strength to support the continuously constructed layer without excessive deformation or damage.

2.5. Strength Testing

To investigate the hardened properties, the flexural and compressive testing have been carried out. The distance between the two support columns in the flexural testing is taken as 100 mm. Flexural strength is calculated according to Equation (2). After the flexural test, two specimens are generated for compressive test with a compressive area of 40 × 40 mm. The arithmetic mean value of the measured flexural strength of three specimens in each group is taken.

$$f_{ts} = \frac{2P}{\pi A} = 0.637\frac{P}{A} \quad (2)$$

The compressive strength is calculated as

$$f_{cc} = \frac{P}{A} \quad (3)$$

where f_{ts} is the splitting tensile strength (MPa), P is failure load (N), A is specimen split surface area (mm^2), and f_{cc} is compressive strength (MPa).

2.6. SEM Imaging

To expose the strength mechanism of UTF-3DPC, the microstructures of concrete are imaged by SEM scanning. The instrument for SEM pictures is JSM-7500F scanning electron microscope produced by Japan Electronics Corporation. The software, Image-Pro Plus 6.0, is used to analyze the porosity of specimens. This software can perform pixel analysis based on the microscopic images.

3. Results and Discussion

By means of the controlled variable method, the optimal mix design of UTF-3DPC is obtained. UTF-3DPCs flowability, fast solidification, strength, and microstructure are evaluated. The precision of 3DPC has been effectively enhanced by ultra-thin filament.

3.1. Flowability

To find out the optimal flowability, the penetration resistance and yield stress of 5 mix ratios of UTF-3DPC have been measured as shown in Figure 5. Comparison of the yield stress of W550SDSC14, W600SDSC14, and W650SDSC14 shows that the yield stress decreases continuously from 387 Pa to 124 Pa as the dosage of the water increases as shown in Figure 5f. Also, as the dosage of SDSC increases continuously from 7 g to 21 g, the yield stress decreases from 248 Pa to 144 Pa as shown in Figure 5f. The yield stress of W600SDSC14 is 185 Pa, which is in the middle of all UTF-3DPC. Generally, the appropriate yield stress ensures UTF-3DPC being extruded continuously and robustly. Extrusion testing proves that 185 Pa is the best yield stress for UTF-3DPC, which is less than 250.2 Pa of 30 mm-rectangular-nozzle 3DPC [5], but it is contained in 66–230 Pa of 20 mm-circular-nozzle 3DPC [30].

From Figure 5f, it can be seen that with the increase in water dosage from 550 to 600, the water dosage only increased by 50, but the yield stress decreased by 31%. As the yield stress decreased, the flowability gradually increased, indicating that with an increase in the water content, the flowability gradually increased, but it is important to note that the flowability may affect the printing quality. The reason is that the incorporation of water makes the adhesion force between the various components of the 3DPC become worse, and it is not so easy to gather into groups.

As a water retaining agent, SDSC can reduce the water loss to a certain extent and promote the flowability, especially when extrusion-based UTF-3DPC slurry is compressed by air pressure. The absence of SDSC always causes water to seep out of UTF-3DPC slurry under the air pressure. The most important function of SDSC is to increase the water retention of concrete and make the slurry be extruded smoothly. Therefore, it is necessary to strictly control the water consumption and the amount of SDSC during the experiment to achieve a balance between the flowability and fast solidification of concrete.

3.2. Fast Solidification

The solidification results of the experiment are shown in Figure 5. The horizontal dashed line in Figure 5a–e represents the penetration resistance of 3.8 MPa, which corresponds to the initial setting time of the printing material. The solidification time curves of five different mix ratios of concrete have been fitted, respectively. It is indicated that they are all approximately in line with the relationship of exponential functions such as $y = Ae^{Bx}$. It can be seen that the amount of water is dominating in the entire solidification testing process.

From Figure 5, it can be seen that as water dosage gradually increases, the slope of the curve gradually decreases, which means that the hardening rate of 3DPC gradually decreases, which is in line with expectations. However, for the increase in the amount of SDSC used, its effect of enhancing early strength and accelerating hydration reactions is not as significant.

The late solidification means that sufficient strength cannot be obtained in the early stage to support the printing of subsequent layers. On the contrary, too early solidification causes the decrease of the adhesive strength between the previous layer and the subsequent layer, which is also an unfavorable factor for constructability. So, either too early or late solidification results in UTF-3DPC tilting or even collapsing during the printing process. The initial setting time of our proposed UTF-3DPC is about 8 h, which is much longer than the solidification time of other larger-filament printers [5].

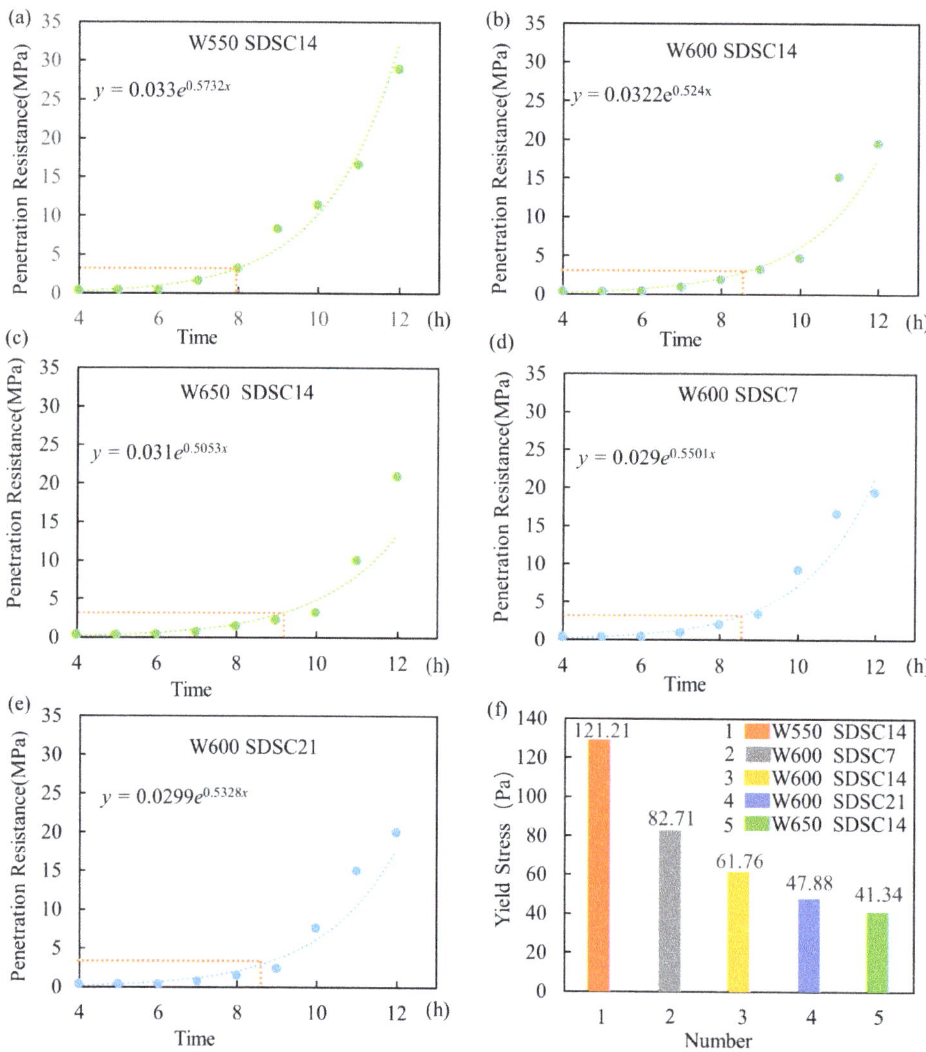

Figure 5. Comparison of fast solidification and flowability data under five different mix ratios. (**a**) W550 means adding water 550 mL, SDSC7 means adding SDSC 7 g, (**b**) W600 means adding water 600 mL, SDSC14 means adding SDSC 14 g, (**c**) W650 means adding water 650 mL, SDSC14 means adding SDSC 14 g, (**d**) W600 means adding water 600 mL, SDSC7 means adding SDSC 7 g, (**e**) W600 means adding water 600 mL, SDSC21 means adding SDSC 21 g. (**f**) The flowability data of printing materials with different mixing ratios.

3.3. Printability Evaluation

Extrudability is a comprehensive manifestation of flowability and rapid solidification. The satisfactory extrudability is manifested by uniform discharge, no clogging, and excessive discharge. When a pressure of 2 bar was applied, the granular slurry cannot be continuously extruded out due to too low air pressure or too dry slurry. Appropriate water, spherical SF and FA, and SDSC are introduced to increase flowability. When 5 bar air pressure is used for extrusion in the experiment process, too much slurry was extruded at one time, which caused the experiment to fail. After reducing the air pressure to 4 bar and increasing the amount of water reducing agent, the extrusion can be performed normally.

It can be seen that many attempts are needed in the process of using the technology of full seal extrusion, and different material properties need different technical parameters to improve the printing quality. Compared with direct extrusion, the construction process of full seal extrusion is more complex, but to some extent, it can improve printing accuracy.

In the experiment, it was found that the printing effect is the best when the yield stress is 50 Pa to 65 Pa through observation and measurement. As can be seen from Figure 6, the relevant data of the fitting curve can be used to obtain the best water consumption and the best dosage of SDSC for the best printing effect. The percentages mentioned here refer to the ratio of SDSC and water to the absolute mass of the cementitious material.

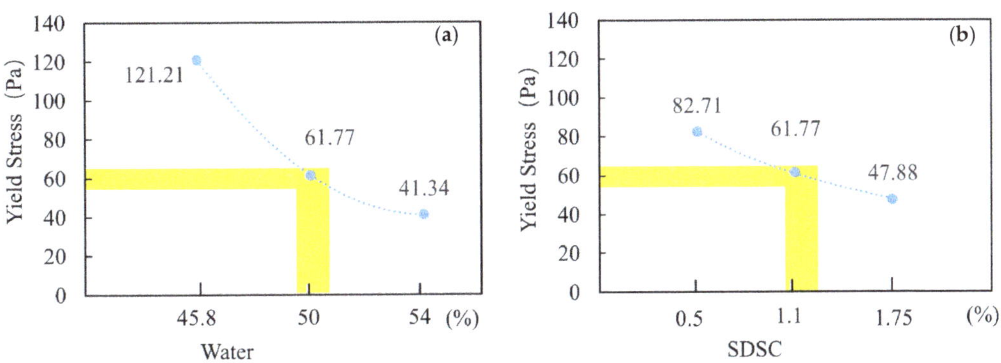

Figure 6. The (**a**) Change in yield stress with changes in water consumption curve and (**b**) change in yield stress with changes in SDSC consumption curve.

In fact, in ordinary 3DPC of larger than 10 mm nozzle, the extrusion is just used to deliver the concrete slurry from material tank to pump hopper, while printing is attained by self-gravity of the slurry and auger force. The extruder of our UTF-3DPC printer does not include pump hopper and the auger, because the pump and auger disturb the sealed slurry, which causes failure of printing process. Many attempts prove that fully sealed extrusion guarantees smooth printing.

3.4. Opening Time and Printing Speed

To obtain optimal printing speed, the printing speeds of 10 mm/s, 30 mm/s, and 50 mm/s were used. Table 3 summarizes the extrudability and buildability parameters of the mixtures printed at different speeds. The successful extrudability and buildability are denoted as 'Y', while the failures are denoted as 'N'.

The main difficulty encountered in the printing process is that extrusion-based 3DPC needs to match the printing speed and extrusion speed; otherwise, cracks, breaks, or accumulations occur. When the slurry is extruded out at 10 mm/s printing speed, the slurry accumulates, while at a speed of 50 mm/s, the slurry bears excessive tension and becomes discontinuous.

Table 3. Evaluation of extrudability and buildability of the mixtures.

Print Speed (mm/s)	W550 SDSC14	W600 SDSC14	W650 SDSC14	W600 SDSC7	W600 SDSC21
10	N/N	Y/N	Y/N	Y/N	Y/N
30	N/N	Y/Y	Y/N	Y/N	Y/Y
50	N/N	Y/N	Y/N	Y/N	Y/N

Note: Y/Y denotes successful extrudability/buildability; Y/N denotes successful extrudability but unsuccessful buildability.

3.5. Strength Analysis

Because the common compressive and flexural directions of concrete are in the Z direction, the specimens of the three different molding methods are all carried out in the Z direction. Under the above mix ratios (Table 1), three types of specimens have been produced, namely printing, mold-cast-vibrating, and mold-cast-non-vibrating. The average compressive strengths of these three in Figure 7a are 20.1 MPa, 31.6 MPa, and 22.35 MPa, respectively. Compared with the mold-cast-vibrating specimens of the same mix ratio, the compressive strengths of the printed and mold-cast-non-vibrating specimens decrease by 36% and 29%, respectively. The average flexural strengths of these three types are 4.54 MPa, 5.98 MPa, and 4.92 MPa, respectively. The printed and mold-cast-non-vibrating specimens are 24% and 17.7% lower than the mold-cast-vibrating specimens. Compressive and flexural strength loss of the printed specimens result from relatively loose bonding between the middle filaments, as shown in Figure 8a, which is mainly due to the fact that 40 mm width of specimens is not exactly divided by 6 mm of filaments.

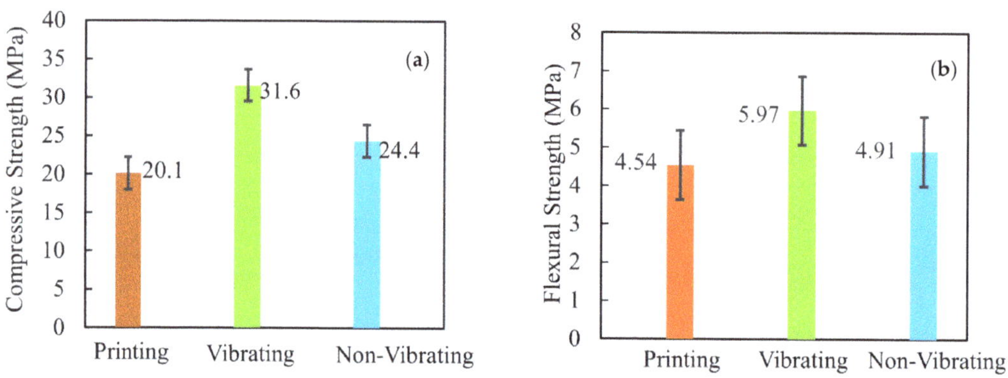

Figure 7. Comparison of compressive (**a**) and flexural (**b**) strengths of specimens with the same mix ratio and different molding methods. *Printing* represents printed specimens. *Vibrating* represents mold-cast-vibrating specimens. *Non-Vibrating* represents mold-cast-non-vibrating specimens.

From the perspectives of flexural strength and compressive strength, there appears to be a significant gap, which is due to the obvious anisotropy of 3DPC, meaning that there are significant differences in strength in the X, Y, and Z directions during strength testing. During flexural strength testing, the combination bending strength of multiple printed filaments is measured until all filaments have broken, at which point the entire specimen is completely destroyed. For compressive strength testing, during the experiment, due to possible gaps between the filaments, individual filaments may crack, at which point the experimental equipment recognizes the damage and stops testing, only measuring a filament of the strength. This is also one of the reasons why printed specimens have significantly lower strengths compared to cast specimens.

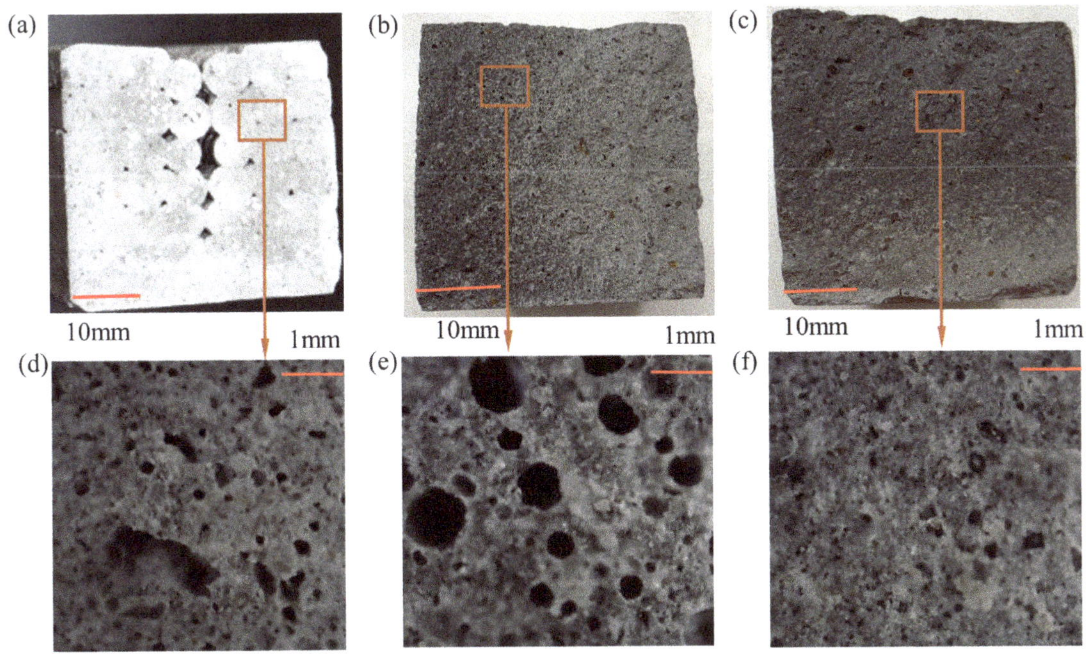

Figure 8. (**a**,**d**) are the cross-section and meso-scale view of the printed specimen. (**b**,**e**) are the cross-section and meso-scale view of the mold-cast-non-vibrating specimen. (**c**,**f**) are cross-section and meso-scale view of mold-cast-vibrating specimen.

3.6. Porosity Analysis

As shown in Figure 8a–c, macro-scale comparison of printing, mold-cast-non-vibrating, and mold-cast-vibrating specimens cross section shows that mold-cast-non-vibrating and mold-cast-vibrating specimens present more uniformly distributed porosity structures than printed ones. Obvious voids occur in the printed specimen. Down to meso-scale level, in Figure 8d,e, printing and vibrating specimens possess smaller pores than non-vibrating, which is proved by the statistics on the total porosity area and proportion in Figure 9a. The largest macro-scale porosity of the printed specimen is due to the fact that 40 mm width of the specimen is not exactly divided by 6 mm width of the filament, which causes inborn void in the middle section. Consequently, the voids weaken the strengths of the printed specimen. The meso-scale picture of the printed specimen is taken in the filament section. Because our UTF-3DPC is compressed dense in the material tank, the compressed slurry is denser than mold-cast-vibrating and mold-cast-non-vibrating. As soon as the voids between filaments are solved, the UTF-3DPC is more promising style than traditional construction.

According to Figure 9b–d, various porosities are evenly distributed in three specimens. But pore sizes of the printed specimen range from 0.1 to 0.4 mm, the mold-cast-non-vibrating 0.04 to 0.8 mm, and the mold-cast-vibrating 0.01–0.08 mm, which also confirms that the strength of the mold-cast-vibrating specimen is the best among these three.

However, it should be noted that during the process of porosity analysis, only a portion of the image is selected for analysis, and the most characteristic image is used to demonstrate the different effects brought about by different forming methods. Differences in other sectional specimens can be further explored through X-ray scanning.

Further down to micro-scale level, comparison between micro-scale images of these three indicates that calcium silicate hydrate (C-S-H) effectively adheres to SF, FA, SDSC, and sand in Figures 10–12. C-S-H, which is a gel-like substance formed after cement hydration reaction, can be well distinguished in micro-images. FA, as a supplementary cementing material, presents as round balls, while SF acts as a lubricant to reduce shear force between various components, which facilitates better extrudability of the material. But the existence of voids and cracks in different sizes and proportions results in the different macro-scale strength of these three specimens. As shown in Figures 10a, 11b and 12b, the printed and the mold-cast-vibrating specimens are denser than the mold-cast-non-vibrating. Due to the compressing action of the piston, the UTF-3DPC has the approximate density as the mold-cast-vibrating specimen. If the bad adhesion between filaments is improved, the strength of the printed specimen will reach the strength of the mold-cast-vibrating specimen.

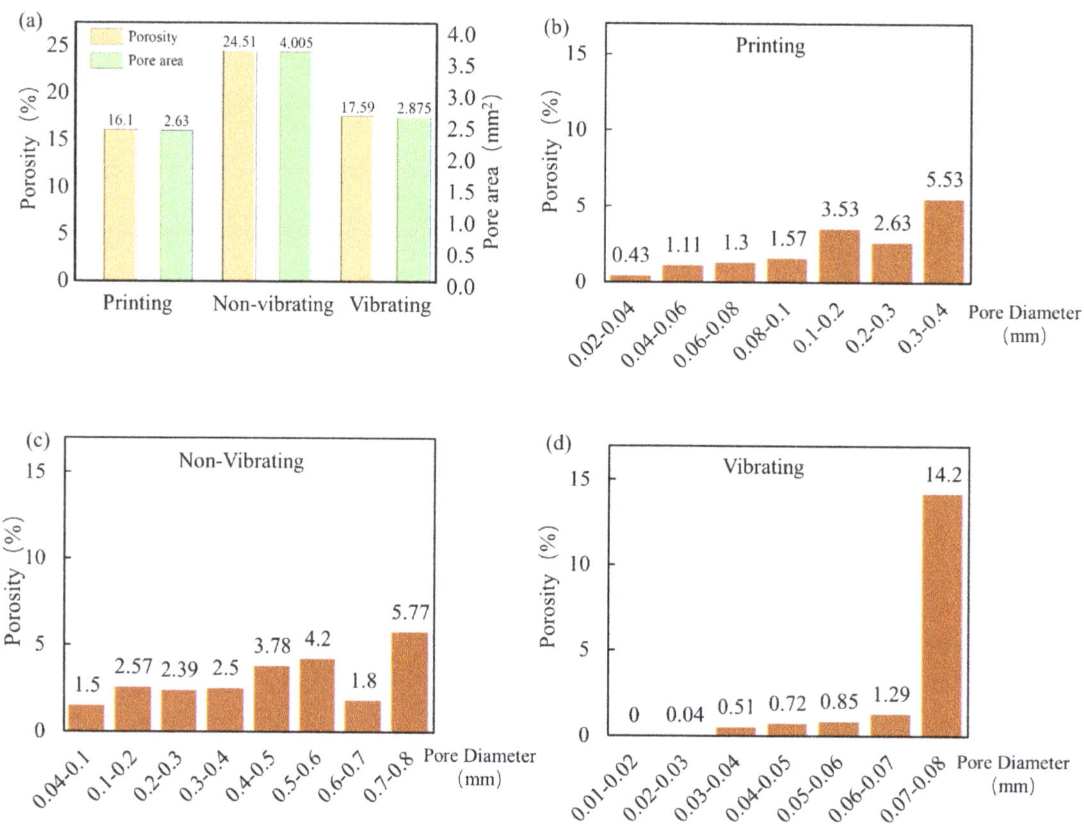

Figure 9. Comparison of porosity of specimens made by three different methods. (**a**) The total porosity and pore area of the specimens. (**b**) is the pore distribution of the printed specimen. (**c**) is the pore distribution of the mold-cast-non-vibrating specimen. (**d**) is the pore distribution of the mold-cast-vibrating specimen.

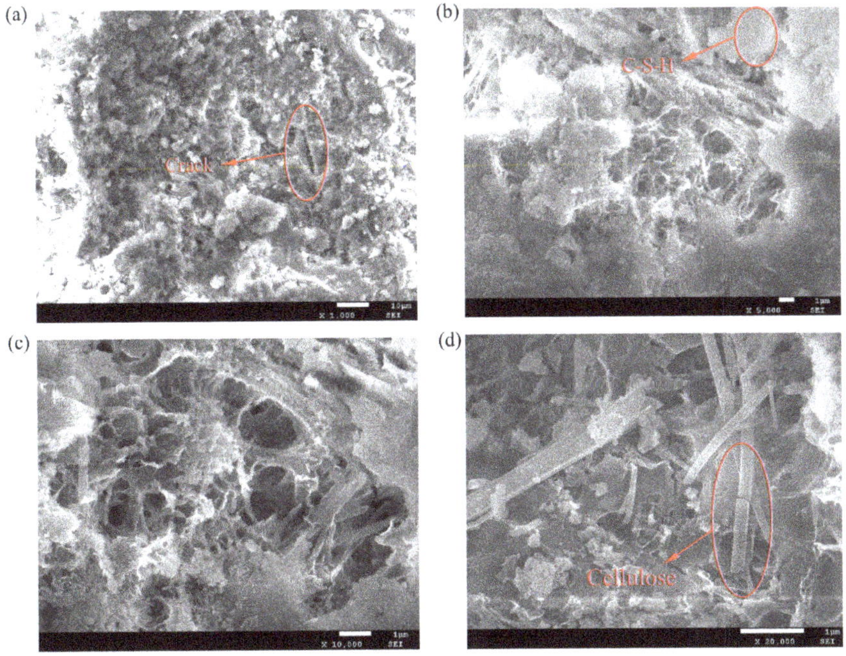

Figure 10. 1000 (**a**), 5000 (**b**), 10,000 (**c**), and 20,000 (**d**) times magnified SEM images of printed specimens.

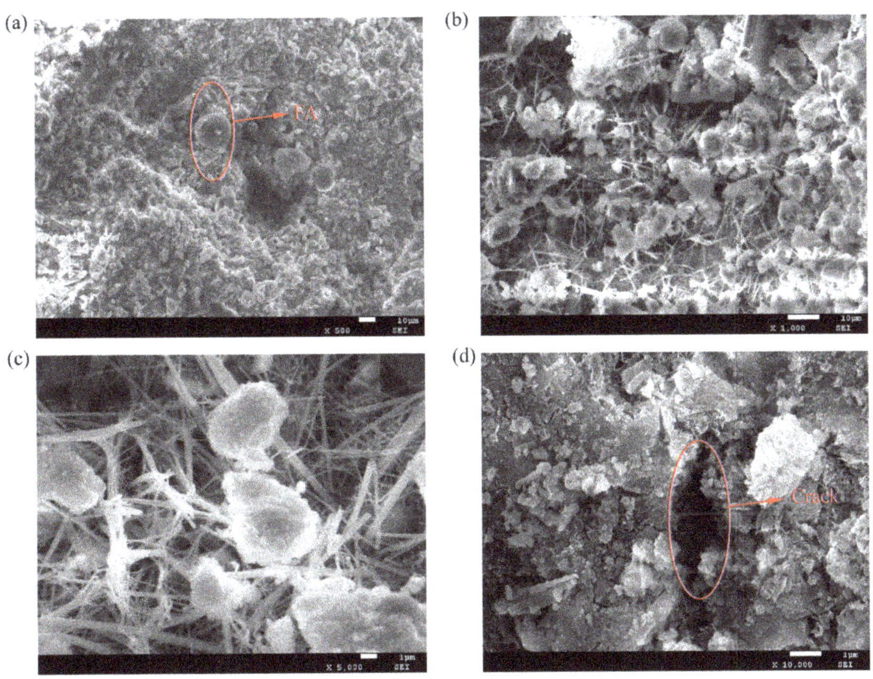

Figure 11. 500 (**a**), 1000 (**b**), 5000 (**c**), and 10,000 (**d**) times magnified SEM images of mold-cast-non-vibrating specimens.

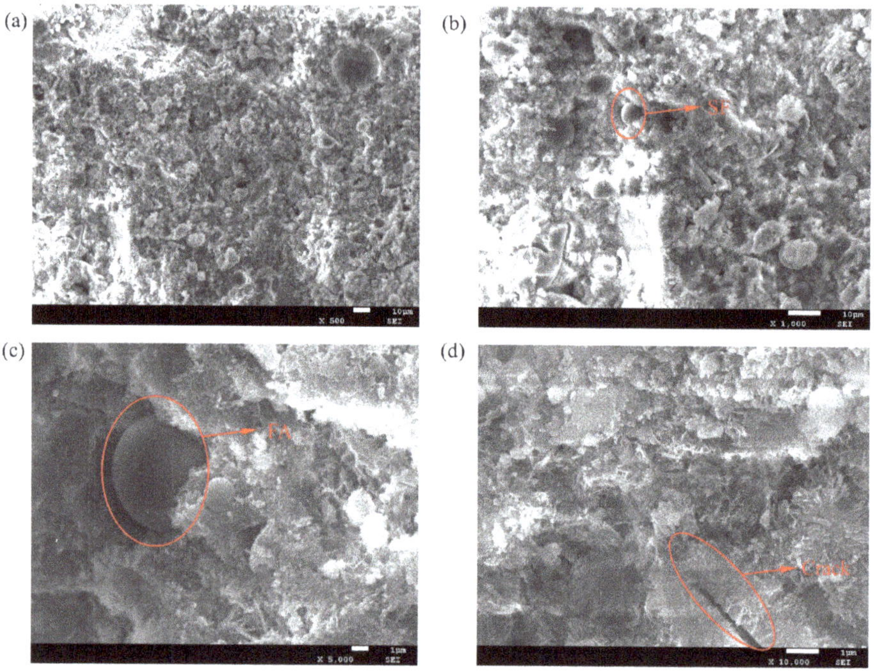

Figure 12. 500 (**a**), 1000 (**b**), 5000 (**c**), and 10,000 (**d**) times SEM images of the mold-cast-vibrating specimens.

4. Conclusions

To improve the precision of 3DPC, the ultra-thin filament 3DPC (UTF-3DPC) is proposed and a customized concrete 3D printer with 6 mm circular nozzle and fully sealed extrusion system is developed. To obtain optimal flowability, extrudability, and fast solidification of UTF-3DPC, FA, SF, SDSC, and water reducer are introduced to mix with OPC, water, and sand in different proportions. To obtain the optimal mix ratio, the yield stress, the penetration volume, and the initial setting time of UTF-3DPC of different mix ratios are measured according to the standard methods. Optimal mix ratio, appropriate printing and extruding speed, and air pressure ensure succeeding the fully sealed UTF-3DPC printing. To verify the practicability of UTF-3DPC, the compressive and tensile strengths of UTF-3DPC are compared with mold-cast-vibrating and mold-cast-non-vibrating specimens. To explore the underlying mechanism of UTF-3DPC strength, the macro-scale, meso-scale, and micro-scale porosity structures are imaged via camera and SEM method. Some conclusions are made as follows:

1. UTF-3DPC can effectively improve print accuracy, but adjustments to printing parameters are significant during the printing process to achieve good print results;
2. The 6 mm-filament 3DPC can be printed successfully at 30 mm/s printing speed, 0.4 MPa air pressure;
3. UTF-3DPC includes FA, SF, SDSC, water reducer, sand, water, and OPC. FA and SF are helpful to optimize the gradation of UTF-3DPC. SDSC plays a key role in improving the flowability, extrudability, and printability, and retaining water under the air pressure;
4. The optimal dynamic yield stress and initial setting time of UTF-3DPC are 175 Pa to 200 Pa, and 8 h, respectively;
5. The strength of UTF-3DPC reaches 80% of mold-cast-vibrating concrete;
6. If the inter-filament adhesion of UTF-3DPC can be improved, the strength of UTF-3DPC will be further enhanced;

7. The meso-scale and micro-scale structures of UTF-3DPC indicate that air pressure is beneficial to enhance its strength due to fully sealed extrusion system. Under the air pressure, fully sealed UTF-3DPC becomes denser than mold-cast-non-vibrating concrete;
8. The findings reported here provide a pathway to improve the precision of 3DPC and extend the application of 3D printing technology in engineering.

Author Contributions: Conceptualization, J.S. and Y.L. (Yangbo Li); methodology, X.Z. and Y.L. (Yujia Li); software, J.S. and Y.L. (Yujia Li); formal analysis, W.L., X.Z. and C.H.; resources, W.L., X.Z., Y.L. (Yangbo Li) and C.H.; writing—original draft preparation, J.S.; writing—review and editing, Y.L. (Yangbo Li); supervision, W.L., Y.L. (Yangbo Li) and C.H.; project administration, W.L. All authors have read and agreed to the published version of the manuscript.

Funding: This research received no external funding.

Data Availability Statement: All relevant data are within the paper.

Conflicts of Interest: The authors declare no conflicts of interest.

List of Abbreviations in the Research

3DPC	three-dimensional printable concrete
UTF-3DPC	ultra-thin-filament three-dimensional printable concrete
FA	Fly ash
SF	Silica fume
OPC	ordinary Portland cement
SDSC	sodium dodecyl sulfate and cellulose
SEM	scanning electron microscope

References

1. Furet, B.; Poullain, P.; Garnier, S. 3D Printing for Construction Based on a Complex Wall of Polymer-Foam and Concrete. *Addit. Manuf.* **2019**, *28*, 58–64. [CrossRef]
2. Hager, I.; Golonka, A.; Putanowicz, R. 3D Printing of Buildings and Building Components as the Future of Sustainable Construction? *Procedia Eng.* **2016**, *151*, 292–299. [CrossRef]
3. Vantyghem, G.; De Corte, W.; Shakour, E.; Amir, O. 3D Printing of a Post-Tensioned Concrete Girder Designed by Topology Optimization. *Autom. Constr.* **2020**, *112*, 103084. [CrossRef]
4. Salet, T.A.M.; Ahmed, Z.Y.; Bos, F.P.; Laagland, H.L.M. Design of a 3D Printed Concrete Bridge by Testing. *Virtual Phys. Prototyp.* **2018**, *13*, 222–236. [CrossRef]
5. Bong, S.H.; Xia, M.; Nematollahi, B.; Shi, C. Ambient Temperature Cured "just-Add-Water" Geopolymer for 3D Concrete Printing Applications. *Cem. Concr. Compos.* **2021**, *121*, 104060. [CrossRef]
6. Ma, G.; Wang, L.; Ju, Y. State-of-the-Art of 3D Printing Technology of Cementitious Material—An Emerging Technique for Construction. *Sci. China-Technol. Sci.* **2018**, *61*, 475–495. [CrossRef]
7. Xiao, J.; Ji, G.; Zhang, Y.; Ma, G.; Mechtcherine, V.; Pan, J.; Wang, L.; Ding, T.; Duan, Z.; Du, S. Large-Scale 3D Printing Concrete Technology: Current Status and Future Opportunities. *Cem. Concr. Compos.* **2021**, *122*, 104115. [CrossRef]
8. Bai, G.; Wang, L.; Wang, F.; Ma, G. In-Process Reinforcing Method: Dual 3D Printing Procedure for Ultra-High Performance Concrete Reinforced Cementitious Composites. *Mater. Lett.* **2021**, *304*, 130594. [CrossRef]
9. Bhattacherjee, S.; Basavaraj, A.S.; Rahul, A.V.; Santhanam, M.; Gettu, R.; Panda, B.; Schlangen, E.; Chen, Y.; Copuroglu, O.; Ma, G.; et al. Sustainable Materials for 3D Concrete Printing. *Cem. Concr. Compos.* **2021**, *122*, 104156. [CrossRef]
10. Arunothayan, A.R.; Nematollahi, B.; Ranade, R.; Khayat, K.H.; Sanjayan, J.G. Digital Fabrication of Eco-Friendly Ultra-High Performance Fiber-Reinforced Concrete. *Cem. Concr. Compos.* **2022**, *125*, 104281. [CrossRef]
11. Bai, G.; Wang, L.; Ma, G.; Sanjayan, J.; Bai, M. 3D Printing Eco-Friendly Concrete Containing under-Utilised and Waste Solids as Aggregates. *Cem. Concr. Compos.* **2021**, *120*, 104037. [CrossRef]
12. Moini, M.; Olek, J.; Youngblood, J.P.; Magee, B.; Zavattieri, P.D. Additive Manufacturing and Performance of Architectured Cement-Based Materials. *Adv. Mater.* **2018**, *30*, 1802123. [CrossRef] [PubMed]
13. Panda, B.; Ruan, S.; Unluer, C.; Tan, M.J. Improving the 3D Printability of High Volume Fly Ash Mixtures via the Use of Nano Attapulgite Clay. *Compos. Part B-Eng.* **2019**, *165*, 75–83. [CrossRef]
14. Arunothayan, A.R.; Nematollahi, B.; Ranade, R.; Bong, S.H.; Sanjayan, J. Development of 3D-Printable Ultra-High Performance Fiber-Reinforced Concrete for Digital Construction. *Constr. Build. Mater.* **2020**, *257*, 119546. [CrossRef]
15. Khelifi, H.; Perrot, A.; Lecompte, T.; Ausias, G. Design of Clay/Cement Mixtures for Extruded Building Products. *Mater. Struct.* **2013**, *46*, 999–1010. [CrossRef]

16. Marchment, T.; Sanjayan, J. Reinforcement Method for 3D Concrete Printing Using Paste-Coated Bar Penetrations. *Autom. Constr.* **2021**, *127*, 103694. [CrossRef]
17. Salman, N.M.; Ma, G.; Ijaz, N.; Wang, L. Importance and Potential of Cellulosic Materials and Derivatives in Extrusion-Based 3D Concrete Printing (3DCP): Prospects and Challenges. *Constr. Build. Mater.* **2021**, *291*, 123281. [CrossRef]
18. Ducoulombier, N.; Mesnil, R.; Carneau, P.; Demont, L.; Bessaies-Bey, H.; Caron, J.-F.; Roussel, N. The "Slugs-Test" for Extrusion-Based Additive Manufacturing: Protocol, Analysis and Practical Limits. *Cem. Concr. Compos.* **2021**, *121*, 104074. [CrossRef]
19. Li, W.; Ma, H.; Li, Z.; Ma, G.; Guan, J. Cementitious Composites Blending with High Belite Sulfoaluminate and Medium-Heat Portland Cements for Largescale 3D Printing. *Addit. Manuf.* **2021**, *46*, 102189. [CrossRef]
20. Ma, G.; Li, Y.; Wang, L.; Zhang, J.; Li, Z. Real-Time Quantification of Fresh and Hardened Mechanical Property for 3D Printing Material by Intellectualization with Piezoelectric Transducers. *Constr. Build. Mater.* **2020**, *241*, 117982. [CrossRef]
21. Wang, L.; Ma, G.; Liu, T.; Buswell, R.; Li, Z. Interlayer Reinforcement of 3D Printed Concrete by the In-Process Deposition of U-Nails. *Cem. Concr. Res.* **2021**, *148*, 106535. [CrossRef]
22. Sanjayan, J.G.; Nematollahi, B.; Xia, M.; Marchment, T. Effect of Surface Moisture on Inter-Layer Strength of 3D Printed Concrete. *Constr. Build. Mater.* **2018**, *172*, 468–475. [CrossRef]
23. Le, T.T.; Austin, S.A.; Lim, S.; Buswell, R.A.; Law, R.; Gibb, A.G.F.; Thorpe, T. Hardened Properties of High-Performance Printing Concrete. *Cem. Concr. Res.* **2012**, *42*, 558–566. [CrossRef]
24. Perrot, A.; Rangeard, D.; Courteille, E. 3D Printing of Earth-Based Materials: Processing Aspects. *Constr. Build. Mater.* **2018**, *172*, 670–676. [CrossRef]
25. Paul, S.C.; Tay, Y.W.D.; Panda, B.; Tan, M.J. Fresh and Hardened Properties of 3D Printable Cementitious Materials for Building and Construction. *Arch. Civ. Mech. Eng.* **2018**, *18*, 311–319. [CrossRef]
26. Lim, J.H.; Panda, B.; Quang-Cuong, P. Improving Flexural Characteristics of 3D Printed Geopolymer Composites with In-Process Steel Cable Reinforcement. *Constr. Build. Mater.* **2018**, *178*, 32–41. [CrossRef]
27. Panda, B.; Paul, S.C.; Mohamed, N.A.N.; Tay, Y.W.D.; Tan, M.J. Measurement of Tensile Bond Strength of 3D Printed Geopolymer Mortar. *Measurement* **2018**, *113*, 108–116. [CrossRef]
28. Ma, G.; Li, Z.; Wang, L. Printable Properties of Cementitious Material Containing Copper Tailings for Extrusion Based 3D Printing. *Constr. Build. Mater.* **2018**, *162*, 613–627. [CrossRef]
29. Ghourchian, S.; Butler, M.; Krueger, M.; Mechtcherine, V. Modelling the Development of Capillary Pressure in Freshly 3D-Printed Concrete Elements. *Cem. Concr. Res.* **2021**, *145*, 106457. [CrossRef]
30. Liu, C.; Wang, X.; Chen, Y.; Zhang, C.; Ma, L.; Deng, Z.; Chen, C.; Zhang, Y.; Pan, J.; Banthia, N. Influence of Hydroxypropyl Methylcellulose and Silica Fume on Stability, Rheological Properties, and Printability of 3D Printing Foam Concrete. *Cem. Concr. Compos.* **2021**, *122*, 104158. [CrossRef]
31. Zhang, N.; Xia, M.; Sanjayan, J. Short-Duration near-Nozzle Mixing for 3D Concrete Printing. *Cem. Concr. Res.* **2022**, *151*, 106616. [CrossRef]
32. Lu, B.; Li, M.; Wong, T.N.; Qian, S. Effect of Printing Parameters on Material Distribution in Spray-Based 3D Concrete Printing (S-3DCP). *Autom. Constr.* **2021**, *124*, 103570. [CrossRef]
33. Mazhoud, B.; Perrot, A.; Picandet, V.; Rangeard, D.; Courteille, E. Underwater 3D Printing of Cement-Based Mortar. *Constr. Build. Mater.* **2019**, *214*, 458–467. [CrossRef]
34. Lowke, D.; Vandenberg, A.; Pierre, A.; Thomas, A.; Kloft, H.; Hack, N. Injection 3D Concrete Printing in a Carrier Liquid-Underlying Physics and Applications to Lightweight Space Frame Structures. *Cem. Concr. Compos.* **2021**, *124*, 104169. [CrossRef]
35. Jiang, X.; Li, Y.; Yang, Z.; Li, Y.; Xiong, B. Harnessing Path Optimization to Enhance the Strength of Three-Dimensional (3D) Printed Concrete. *Buildings* **2024**, *14*, 455. [CrossRef]
36. *GB/T 2419-2005;* Test Method for Fluidity of Cement Mortar. Chinese National Testing Standard: Beijing, China, 2005.

materials

Article

Properties of Fiber-Reinforced Geopolymer Mortar Using Coal Gangue and Aeolian Sand

Yupeng Li [1], Akelamjiang Maimait [1,2], Jianjun Cheng [1,*], Yanfu Duan [1], Yuanqing Chen [1] and Hongguang Dong [1]

1 College of Water Conservancy & Architectural Engineering, Shihezi University, Shihezi 832003, China; 11774786383@163.com (Y.L.); akelamjiang@126.com (A.M.); lttdyf@163.com (Y.D.); chenyq202302@163.com (Y.C.); dhg19990114@163.com (H.D.)

2 Xinjiang Kusha Highway Development Limited Liability Company, Aksu 843000, China

* Correspondence: chengjianjun@shzu.edu.cn

Abstract: Geopolymers, as a novel cementitious material, exhibit typical brittle failure characteristics under stress. To mitigate this brittleness, fibers can be incorporated to enhance toughness. This study investigates the effects of varying polypropylene fiber (PPF) content and fiber length on the flowability, mechanical properties, and flexural toughness of coal gangue-based geopolymers. Microstructural changes and porosity variations within the Fiber-Reinforced Geopolymer Mortar(GMPF) matrix were observed using scanning electron microscope (SEM) and Low field NMR(LF-NMR) to elucidate the toughening mechanism of PPF-reinforced geopolymers. The introduction of fibers into the geopolymer matrix demonstrated an initial bridging effect in the viscous geopolymer slurry, with a 3.0 vol% fiber content reducing fluidity by 5.6%. Early mechanical properties of GMPF were enhanced with fiber addition; at 1.5 vol% fiber content and 15 mm length, the 3-day flexural and compressive strengths increased by 30.81% and 17.4%, respectively. Furthermore, polypropylene fibers significantly improved the matrix's flexural toughness, which showed an increasing trend with higher fiber content. At a 3.0 vol% fiber content, the flexural toughness index increased by 198.35%. The data indicated that a fiber length of 12 mm yielded the best toughening effect, with an 84.03% increase in the flexural toughness index. SEM observations revealed a strong interfacial bond between fibers and the matrix, with noticeable damage on the fiber surface due to frictional forces, and fiber pull-out being the predominant failure mode. Porosity testing results indicated that fiber incorporation substantially improved the internal pore structure of the matrix, reducing the median pore diameter of mesopores and converting mesopores to micropores. Additionally, the number of harmless and less harmful pores increased by 23.01%, while the number of more harmful pores decreased by 30.43%.

Keywords: geopolymer; polypropylene fiber; mechanical properties; flexural toughness; microscopic analysis

1. Introduction

With the accelerated pace of urbanization, the production of ordinary Portland cement (OPC) has reached 4 billion metric tons and is expected to further increase in the next decade. However, traditional cement production not only consumes vast amounts of raw materials but also releases significant amounts of carbon dioxide during the calcination of limestone, causing long-term adverse environmental impacts. To address this challenge, researchers have explored the use of alkali-activated industrial solid wastes, including fly ash (FA), silica fume (SF), ground granulated blast furnace slag (GGBS), coal gangue (CG), and silica fume, to produce geopolymer concrete. Although geopolymer concrete also exhibits issues such as high brittleness and low tensile strength, it offers several advantages over OPC concrete, including a wider range of material sources, lower carbon emissions, higher early strength, and excellent durability.

The mechanical properties of geopolymers are very similar to those of cement, exhibiting excellent compressive strength and relatively poor tensile strength, which results in a certain degree of brittleness. To improve this brittleness, small fibers are typically randomly mixed into traditional concrete to form what is known as fiber-reinforced concrete (FRC). In common brittle materials, once cracks initiate, they tend to propagate rapidly, leading to a sharp decline in the load-bearing capacity of the matrix. However, in FRC, the fibers distributed within the concrete matrix can intercept crack propagation, slowing down or even stopping crack formation. This phenomenon, known as the crack bridging effect, significantly enhances the toughness of the concrete and maintains its load-bearing capacity even after the initial crack appears. Inspired by this approach, researchers have attempted to incorporate fibers into geopolymers to improve their brittleness, achieving varying degrees of success. Table 1 summarizes recent studies on fiber-reinforced geopolymers, providing specific details on their findings.

Table 1. Research progress of fiber geopolymers in recent years.

Study	Fiber Type	Fiber Content	Fiber Length	Geopolymer Matrix	Key Findings
Li et al. [1]	Steel fiber	0.5 vol%	13 mm	GBFS-based	Steel fibers are less susceptible to corrosion, while effectively enhancing both compressive and flexural strength.
Zhang et al. [2]	Carbon fiber	0.6 vol%	6 mm	Fireclay-based	Carbon fibers with a length of 6 mm demonstrated better crack initiation toughness and resistance to unstable cracking.
Zhang et al. [3]	Polypropylene fiber	1.0 vol%	35 mm	FA- and GBFS-based	Polypropylene (PP) fibers exacerbated the adverse effects of frost heave stress and thermal stress on the geopolymer matrix (GPM), leading to a reduction in frost resistance.
Li et al. [4]	Polyethylene fiber	1.5 vol%	12 mm	FA-, SF-, and GBFS-based	High fiber content of polyethylene fiber (PEF) significantly enhanced the compressive strength and fracture toughness of the specimens.
Zhang et al. [5]	Polypropylene fiber and polyethylene fiber	2.0 vol%	12 mm	FA- and GBFS-based	The addition of PP fibers reduced the tensile strength and crack control ability. However, with a PP fiber replacement rate of 50%, the highest tensile strain capacity was achieved (9.66%).
Zhao et al. [6]	Sisal fiber	1.0 vol%	20 mm	GBFS-based	The addition of SF reduced the density and electrical conductivity (EC), while simultaneously increasing the porosity, long-term flexural strength, and compressive strength.
Liu et al. [7]	Glass fiber	0.6 vol%	9 mm	FA-based	Glass fibers significantly improved the splitting tensile strength and enhanced the pore characteristics, contributing to a more uniform pore distribution.

Intarabut et al. [8] investigated the use of basalt, steel, and polypropylene fibers to enhance fly ash-based geopolymers and found that each fiber type exhibited different failure modes. Steel and polypropylene fibers demonstrated complete bond-slip responses due to their higher ductility, whereas basalt fibers, being brittle, failed through fiber rupture without slip during the pull-out response. Piti [9] addressed the high flexibility but low stiffness of polypropylene fibers by hybridizing them with steel fibers. The results showed that hybridization with steel fibers could improve the flexural performance, toughness, and residual strength of polypropylene fiber-reinforced geopolymers to varying degrees. Ömer [10] studied the effects of polypropylene, steel, and polyamide fibers on the mechanical and durability properties of fly ash-based geopolymers. Experimental results indicated that fiber addition enhanced the mechanical properties, with polypropylene and steel fibers significantly increasing the flexural toughness, while polyamide fibers showed a less pronounced effect. Jiratchaya [11] discovered that incorporating bamboo and glass fibers into metakaolin-based geopolymers effectively enhanced flexural strength. Compared to glass fibers, the rough bamboo fibers formed stronger interlocking structures with the matrix, exhibiting superior bonding capability and increasing flexural strength by 1.9 times with 1% bamboo fiber addition. Zhang Peng et al. [12] found that polypropylene fibers were most effective in enhancing the toughness of fly ash-based geopolymers, achieving a flexural strength of 21.1 MPa at 28 days with a 1.2% fiber content. Mohseni [13] reported a 28% increase in flexural strength with the addition of PF to geopolymer samples. Al-Mashhadani et al. [14] examined the reinforcing effects of polypropylene, steel, and polyvinyl alcohol fibers on fly ash-based geopolymers, concluding that fiber addition did not significantly increase the compressive strength of the mortar. Steel and polyvinyl alcohol fibers increased compressive strength by only 3.37% and 4.26%, respectively, but flexural strength showed significant improvements compared to the control mortar. Guo and Pan [15] found that fiber volume fractions exceeding 0.2% negatively impacted flexural strength, recommending an optimal fiber usage rate of 0.2% for fly ash slag-based geopolymers.

In summary, while research on using fibers to enhance the mechanical properties of cement-based materials has yielded substantial results, studies on the modification of new cementitious materials like geopolymers with fibers remain insufficient. Most research focuses on fly ash-based geopolymers, with few studies examining fiber-reinforced coal gangue-based geopolymers. Therefore, this study explores the effects of different fiber contents and lengths on the workability and mechanical properties of polypropylene fiber-reinforced coal gangue-slag-based geopolymers (GMPF). Additionally, SEM and LF-NMR microscopic characterization methods are used to observe the microstructure within the matrix and investigate the mechanisms by which fibers affect flexural toughness. This research aims to elucidate the mechanical and toughness enhancement effects of polypropylene fibers on coal gangue-based geopolymers, providing references for the application of fiber-reinforced geopolymer composites and the recycling of coal gangue.

2. Materials and Methods

2.1. Raw Materials

(1) Binder Materials and Aggregates

The preparation of geopolymers primarily utilizes two raw materials: coal gangue powder (CG) and ground granulated blast furnace slag (GBFS). Coal gangue powder is a mature industrial product obtained by crushing, ball milling, and high-temperature treatment in a calcination furnace of raw coal gangue. The GBFS is sourced from an industrial park in Northern Xinjiang, China, and is gray-white. According to the GB/T 18046-2017 standard [16], this slag falls under the S75 grade, meeting the standards for use in cement, mortar, and concrete. For mortar preparation, a mixture of aeolian sand (AS) and river sand (RS) is used as aggregates. This combination leverages the hardness and particle shape of aeolian sand with the smoothness and particle size distribution of river sand, aiming to achieve better workability and mechanical strength. The physical and chemical properties of the materials are shown below, and the elemental composition of the raw materials is analyzed by X-ray fluorescence (XRF), as presented in Table 2. Figure 1 illustrates the X-ray diffraction (XRD) and scanning electron microscopy (SEM) analyses of the raw materials. XRD analysis reveals that coal gangue and aeolian sand are mainly composed of quartz, sodium feldspar, and mica phases, while the XRD of slag mainly exhibits a broad hump shape, indicative of an amorphous phase. The particle size analysis of aeolian sand and particle gradation of river sand are shown in Figure 2. The fineness modulus of river sand is 2.77, complying with standard ASTM C33-18 specifications [17].

Table 2. Chemical composition analysis of raw materials.

Materials	SiO_2	Al_2O_3	CaO	Fe_2O_3	Mg_2O_3	Na_2O	K_2O	Other
CG	53.45	43.46	0.317	0.781	0.11	0.17	0.291	1.421
GBFS	32.85	8.62	45.16	0.63	5.69	1.08	0.497	5.473
AS	59.3	11	8.42	9.125	1.52	3.76	3.45	3.425

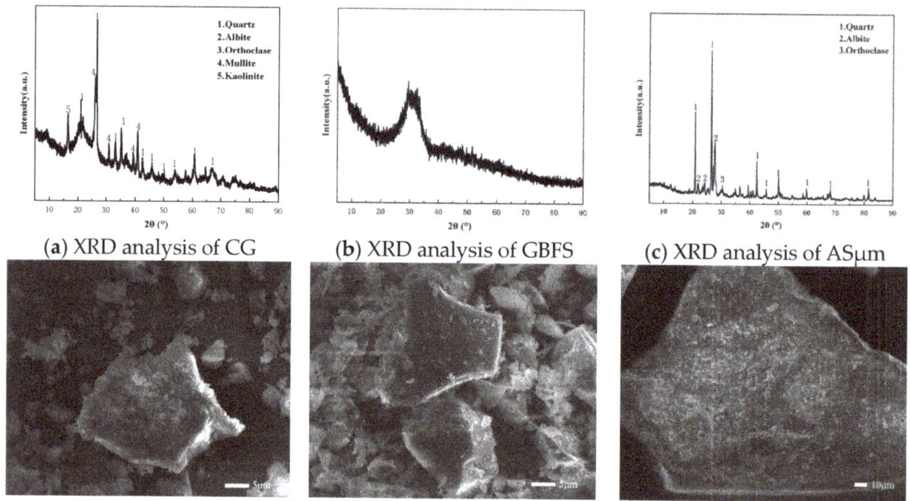

(**a**) XRD analysis of CG (**b**) XRD analysis of GBFS (**c**) XRD analysis of ASμm

(**d**) SEM analysis of CG (**e**) SEM analysis of GBFS (**f**) SEM analysis of AS

Figure 1. XRD and SEM analyses of raw materials.

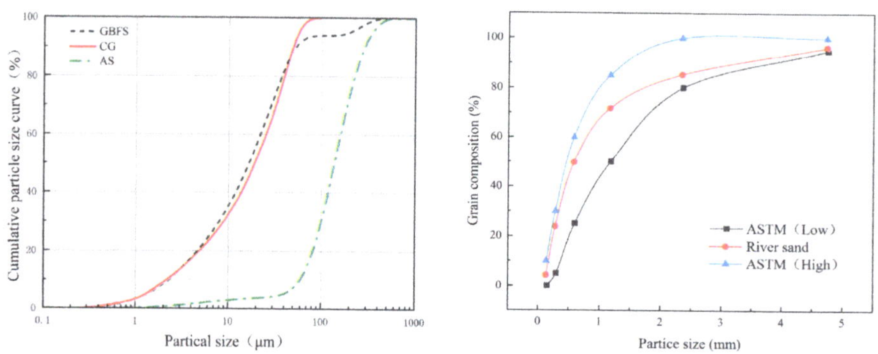

(**a**) CG, GBFS, and AS particle size analysis (**b**) Grain gradation of river sand

Figure 2. Particle size analysis and particle grading.

(2) Alkali Activator

The alkaline activator is prepared by mixing liquid sodium silicate and sodium hydroxide. The original liquid sodium silicate has a modulus (Na_2O/SiO_2) of 3.28, with a Na_2O content of $\geq$8.35% and a SiO_2 content of $\geq$26.54%. Sodium hydroxide is of analytical grade, with a purity of $\geq$98%, and the water used in the experiment is laboratory-grade purified water. To prepare the activator, the NaOH and Na_2SiO_3 solutions are thoroughly stirred, then covered and allowed to age at room temperature for 24 h. Once completely cooled, the solution is ready for use.

(3) Polypropylene Fiber

The main physical and mechanical properties of polypropylene fibers are presented in Table 3.

Table 3. Principal physical and mechanical properties of polypropylene fibers.

Equivalent Diameter (μm)	Tensile Strength (MPa)	Initial Modulus (MPa)	Elongation at Break (%)	Density (g/cm^3)
38	550	4.7×10^3	85	0.91

2.2. Experiment Design

The mix parameters listed below are original contributions by the author. Prior to conducting this experiment, the baseline mix parameters for geopolymer mortar were determined through preliminary testing. The mix proportions for preparing fiber-reinforced geopolymer matrices are provided in Table 4. To prepare the specimens, coal gangue and slag powders are first mixed with the corresponding masses of aeolian sand and river sand aggregates and stirred for 2 min. Once mixing is complete, the pre-prepared alkaline activator solution and water are added rapidly and stirred for 3–5 min. After observing uniform mixing of the slurry, the geopolymer precursor is obtained. The precursor is then slowly poured into molds with dimensions of 40 mm × 40 mm × 160 mm, compacted, covered, and allowed to cure for 24 h. After demolding, the specimens are placed in a standard curing chamber at a temperature of 20 ± 2 °C and a humidity of 95% until the specified age for testing. The specific experimental procedures are outlined in Table 5.

Table 4. Base mix ratio of geopolymer preparation (g).

Coal Gangue	GBFS	Water	Na_2SiO_3	NaOH	Aeolian Sand	River Sand
270.00	270.00	142.83	198.09	37.09	540.00	540.00

Table 5. Experimental protocol for fiber-reinforced geopolymer.

Specimen No.	Coal Gangue (g)	GBFS(g)	Water(g)	Alkali Activator (g)	Fiber Content (g)	Fiber Length (mm)
GMPF0-0	270.00	270.00	142.83	235.18	0	0
GMPF0.5-15	270.00	270.00	142.83	235.18	1.17	15
GMPF1.0-15	270.00	270.00	142.83	235.18	2.33	15
GMPF1.5-15	270.00	270.00	142.83	235.18	3.50	15
GMPF2.0-15	270.00	270.00	142.83	235.18	4.66	15
GMPF3.0-15	270.00	270.00	142.83	235.18	6.99	15
GMPF1.0-6	270.00	270.00	142.83	235.18	2.33	6
GMPF1.0-9	270.00	270.00	142.83	235.18	2.33	9
GMPF1.0-12	270.00	270.00	142.83	235.18	2.33	12
GMPF1.0-15	270.00	270.00	142.83	235.18	2.33	15
GMPF1.0-19	270.00	270.00	142.83	235.18	2.33	19

2.3. Testing Methods

2.3.1. Mechanical Performance Testing and Flowability Testing

Flowability testing follows the recommended method outlined in GB/T 2419-2005 "Test methods for fluidity of cement mortar" [18]. Initially, the mortar is poured into the truncated cone mold in two batches. Subsequently, the truncated cone mold is quickly lifted, allowing the mortar to flow freely on the flow table. The flow table is then activated to vibrate 25 times at a frequency of once per second. After the completion of the vibration, the diameters in two perpendicular directions are measured using calipers, and the average diameter is taken as the flowability result for the mortar batch.

After curing for 3, 7, and 28 days, flexural and compressive strength tests are conducted using a hydraulic servo-controlled universal testing machine. Three prism specimens are taken for flexural strength testing, and the average of three test results is recorded. During flexural strength testing, the loading rate is 1 mm/min until contact with the specimen surface is made, then it increases to 40 N/s. For compressive strength testing, the broken halves of the specimens are placed in fixtures, and the loading rate is set at 2.4 kN/s. The final compressive strength result is the average of six test results.

2.3.2. Flexural Toughness Testing

Flexural toughness is evaluated using the Japanese Society of Civil Engineers (JSCE) SF4 assessment method. The flexural toughness and flexural toughness coefficient are calculated according to the following methods.

The method for calculating flexural toughness is as follows:

$$T_b = \int p(\delta)d_\delta \tag{1}$$

where T_b is the flexural toughness, which is the area under the load–displacement curve at the corresponding deflection. $P(\delta)$ is the load–displacement curve during the three-point bending test.

The method for calculating the flexural toughness coefficient is as follows:

$$\sigma_b = \frac{T_b}{\delta_{tb}} \times \frac{L}{bh^2} \tag{2}$$

where σ_b is the flexural toughness coefficient; T_b is the flexural toughness; δ_{tb} is the deflection at 1/150 L; and L, b, and h are the length, width, and height of the specimen, respectively.

2.3.3. Scanning Electron Microscopy (SEM) and Low-Field Nuclear Magnetic Resonance (LF-NMR) Analysis

Samples, after 28 days, are soaked in anhydrous ethanol for 24 h to halt hydration reactions. Subsequently, the samples are dried to a constant weight. The central portion of the remaining samples is then crushed into fragments approximately 5 mm in size for scanning electron microscopy (SEM) analysis. Half of the fractured specimens are used for volume measurement and then subjected to low-field nuclear magnetic resonance (LF-NMR) analysis. The internal microstructure of the samples is observed using a Japanese JSM IT-800 field emission scanning electron microscope (JEOL, Tokyo, Japan) under an acceleration voltage of 15 kV. The China NIUMAG MacroMR large-aperture nuclear magnetic resonance imaging analysis system (MacroMR02-110H-D, Suzhzhou Niumag Analytical Instruments Corporation, Suzhou, China) is utilized to analyze the internal porosity, pore size distribution, and pore throat distribution of the samples. The magnetic field strength is set to 0.3 T, and the frequency field is set to 23 MHz.

3. Results

3.1. *Flowability*

The impact of different fiber dosages on flow performance is illustrated in Figure 3a. The dosage of 15 mm polypropylene fibers increases from 0 to 3 vol%, leading to a gradual decrease in flow performance. When the dosage reaches 3 vol%, the flow performance is at its lowest, with a decrease of 9.6% compared to the unfibered state. It can be observed from the graph that within the range of 0–1.5 vol% fiber dosage, the decrease in flowability is relatively moderate, ranging from 0.85% to 2.56%, compared to the reference group. However, when the fiber dosage exceeds 1.5 vol%, the decrease in flow performance accelerates noticeably, with flowability decreasing by 2.24% and 5.6% for the GMPF2.0-15 and GMPF3.0-15 groups, respectively, compared to the reference group. Figure 3b illustrates the effect of different fiber lengths on flow performance with a fiber dosage of 1 vol%. From the GMPF0-0 group to the GMPF1.0-9 group, there is a slight improvement in flow performance. When the fiber length is 9 mm, the flowability of the GMPF slurry increases by 2.4% compared to the unfibered state. However, with fiber lengths of 12 mm, 15 mm, and 19 mm, flowability gradually decreases. The GMPF1.0-19 group exhibits the lowest flowability, with a decrease of 5.8% compared to the GMPF1.0-9 group. Overall, the effect of different fiber lengths on flow performance is relatively minor when the fiber dosage is less than 1.5 vol%. This is mainly due to the hydrophobic nature of polypropylene fibers, which typically exhibit low water absorption and are not easily wetted by water. Consequently, water droplets form on the fiber surface and quickly slide off, providing lubrication in the system [19]. Additionally, short fibers in the matrix can fully encapsulate the remaining water to wrap the aggregates, thereby enhancing flow performance. However, for long fibers, more polymer matrix adheres to the surface, demonstrating a preliminary bridging effect. This effect is more significant for the inherently viscous geopolymer precursor, leading to a 3.8% decrease in flow performance compared to the reference group for the GMPF1.0-19 group.

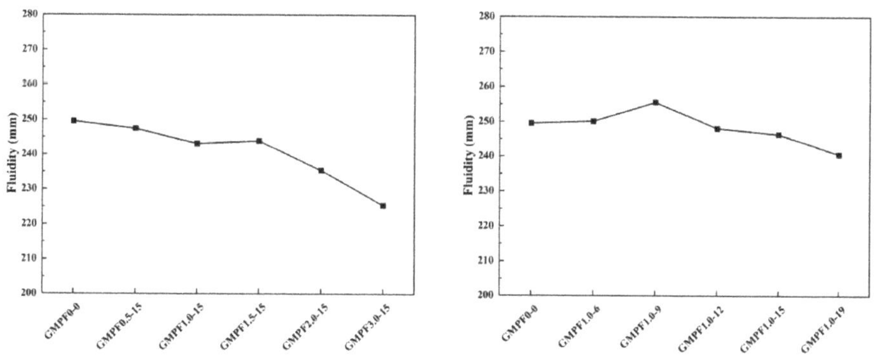

(**a**) Fiber contents (**b**) Fiber lengths

Figure 3. Influence of different fiber contents and fiber lengths on flowability.

3.2. Mechanical Performance

3.2.1. Influence of Different Fiber Contents on Mechanical Strength

Figure 4a presents the flexural strength results at 3d, 7d, and 28d for different fiber dosages. It can be observed that the inclusion of polypropylene fibers leads to a significant improvement in the early flexural strength of the GMPF specimens. The GMPF1.5-15 group achieves a flexural strength of 4.55 MPa at 3d, representing a 30.81% increase compared to the GMPF0-0 group. The GMPF2.0-15 group exhibits the highest flexural strength at 7d and 28d, reaching 5.1 MPa and 6.1 MPa, respectively, corresponding to increases of 24.33% and 11.53% compared to the reference group. As the curing period increases, the efficiency of polypropylene fibers in enhancing the flexural strength of geopolymer matrix gradually diminishes. Comparisons among different fiber dosage groups reveal a trend of initially increasing and then decreasing flexural strength with increasing fiber dosage. The highest flexural strength is achieved at a fiber dosage of 2.0 vol%, with a 25.65% increase compared to the reference group. When the fiber dosage is between 0.5 vol% and 1.0 vol%, the improvement in flexural strength at 28d is extremely limited. Additionally, the flexural strength of the GMPF samples with a 3.0 vol% polypropylene fiber dosage shows a decreasing trend, with a 7.1% decrease compared to the preceding group. Xu et al. [20] observed a similar phenomenon with basalt fiber in fly ash-based geopolymer, where an excessive dosage of basalt fiber adversely affected the flexural strength.

Figure 4b demonstrates the effect of different fiber dosages on compressive strength. At 3d and 7d, the inclusion of fibers positively contributes to the growth of compressive strength. The results indicate that at 3d, the GMPF0.5-15 sample achieves the highest compressive strength of 23.46 MPa, compared to 14.01 MPa for the control sample. At 28d, the sample with a 1.5 vol% polypropylene dosage exhibits the highest compressive strength, with a 17.4% increase compared to the control group. The data suggest that when the polypropylene fiber dosage exceeds 1.5 vol%, further increases result in a decreasing trend in compressive strength. Shaikh et al. [21] found a decrease in compressive strength with PE fibers in geopolymer; however, unlike the substantial decrease in strength observed with PE fibers, this study indicates that PP fibers do not cause a cliff-like drop in matrix compressive strength. It is well known that polypropylene fibers are primarily randomly distributed within the matrix. When the matrix is subjected to tensile stress, the high modulus of elasticity of the fibers enables them to both disperse stress and bear some of the load, thereby enhancing the flexural and compressive strength of the sample. On the other hand, increasing the fiber dosage introduces more bonding interfaces and micro-pores, which affect the polymerization reaction of the geopolymer. Additionally, excessive fiber incorporation will disrupt the uniform distribution of fibers within the matrix, adversely affecting the mechanical strength of the matrix.

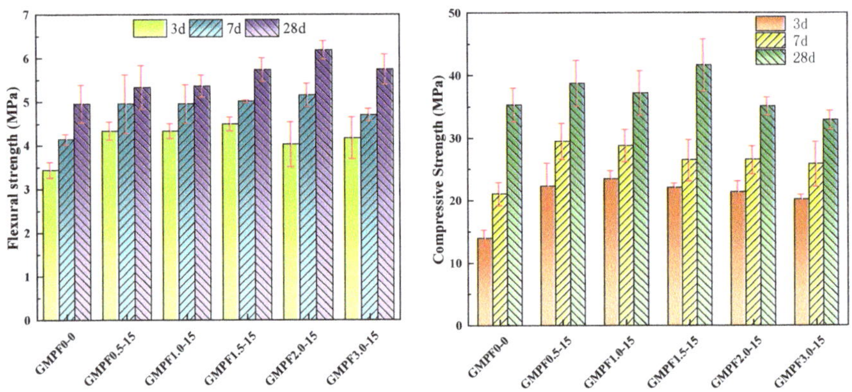

(**a**) Flexural strength results (**b**) Compressive strength results

Figure 4. Mechanical strength results with different fiber contents.

3.2.2. Influence of Different Fiber Lengths on Mechanical Performance

Figure 5a illustrates the flexural strength results for different fiber lengths. Similar to the previous section, the enhancement of early flexural strength by 1 vol% polypropylene fibers is remarkable. The GMPF1.0-6 group achieves the highest flexural strength at 28d, reaching 6.06 MPa, representing a 22.17% increase compared to the reference group. Furthermore, the data indicate that longer fibers do not necessarily result in higher flexural strength. When the fiber length ranges from 9 mm to 19 mm, the flexural strength fluctuates around 5.5 MPa, showing an 11.1% increase compared to the reference group.

Figure 5b depicts the influence of fiber length on compressive strength. The results indicate that fibers of various lengths positively affect compressive strength at 3d and 7d. At 7d, the compressive strength with 15 mm polypropylene fibers is 41.8% higher than that without fibers. The enhancement in flexural strength diminishes with increasing curing time, with the difference in compressive strength between fiber-reinforced samples and the reference group being relatively small at 28d, with a 5.31% increase observed in the GMPF1.0-15 group. This characteristic is similar to the conclusion drawn for polypropylene fibers in cementitious materials, where the influence of fiber length on compressive strength is weak [22]. The experimental results suggest that 6 mm long polypropylene fibers are more suitable for coal gangue-based geopolymer. It was found during mixing that short fibers disperse more uniformly in the geopolymer slurry and can form a better interlocking structure with the geopolymer matrix, exhibiting stronger mechanical properties. Additionally, polypropylene fibers can partially inhibit crack formation in the matrix under compression, thereby enhancing compressive strength. However, due to their lower modulus of elasticity, they cannot prevent cracks from continuing to propagate once they appear, resulting in a weaker enhancement in compressive strength.

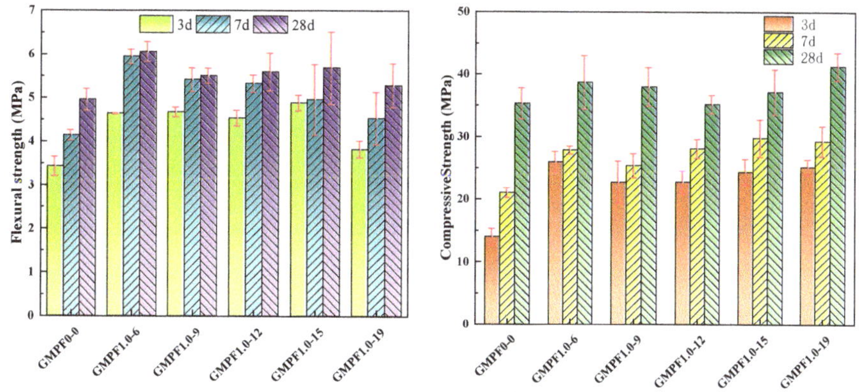

(**a**) Flexural strength results (**b**) Compressive strength results

Figure 5. Mechanical strength results with different fiber lengths.

3.3. Flexural Toughness

3.3.1. Load–Deflection Curve

The bending load–deflection curves for specimens with different fiber contents are presented in Figure 6a. The data indicate a gradual increase in both maximum bending force and maximum deflection with increasing fiber content. It is observed that at a polypropylene fiber content of 2.0 vol%, the maximum bending force increases by 29.8%, and the maximum deflection increases by 36.47%. Additionally, particular attention should be paid to the change in curve morphology at a polypropylene fiber content of 3.0 vol%. In the GMPF3.0-15 group, prior to reaching the maximum bending force, the load–deflection curve is nearly linear, with a smaller slope, and the peak at the maximum bending force is smoother, indicating that the specimen absorbs significantly more energy before failure compared to the other groups, highlighting the most pronounced improvement effect of fibers on the matrix.

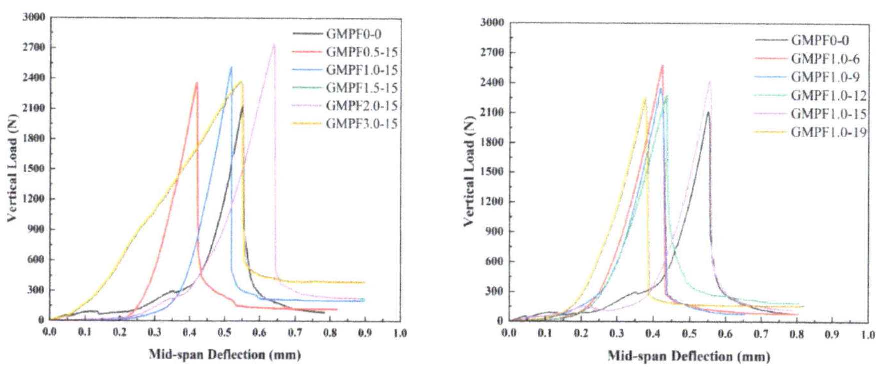

(**a**) Various fiber contents (**b**) Different fiber lengths

Figure 6. Load-deflection curves of GMPF specimens under bending.

The bending load–deflection curves for different fiber lengths are illustrated in Figure 6b. The change trends in the curves are relatively consistent as the fiber length increases from 6 mm to 19 mm. The load–deflection curve can be roughly divided into three stages: initially, when the specimen begins to bend, the testing rate is slow, and the curve exhibits the "cushioning zone", with relatively low stress but considerable displacement; subsequently, the bending force rapidly increases while the displacement changes only slightly, indicating the elastic stage of bending; finally, when reaching the maximum bending force, cracks appear, the bending force drops rapidly, and the displacement continues to increase, entering the strain-softening stage, leading to specimen failure. The experimental results demonstrate that even with an increase in fiber content from 0 to 3 vol% and fiber length from 6 mm to 19 mm, the GMPF composite material does not exhibit strain-hardening characteristics similar to fiber cementitious materials; its load–deflection curve displays typical brittle failure characteristics. Studies by Lin et al. [23]. suggest that compared to steel and polyoxymethylene fibers, polypropylene fibers have poorer bonding ability with the matrix and limited fiber bridging capacity, which may explain the insignificant change in the failure mode.

In summary, the incorporation of polypropylene fibers into coal gangue-based geopolymer composite materials has a beneficial effect on flexural performance. The fibers prevent crack propagation within the geopolymer mortar phase through bridging effects, resist stress applied to the specimens, and transmit stress to regions of the mortar with stronger resistance to bending [24]. Furthermore, the composite material absorbs more energy when subjected to stress. Once cracks appear in the specimen, the fibers fracture and debond from the matrix one by one, dissipating energy in the process, effectively enhancing the material's flexural strength.

3.3.2. Influence of Fibers on Flexural Toughness Coefficient

The flexural toughness was assessed using the Japanese Society of Civil Engineers (JSCE) SF4 evaluation method, as described in Section 2.3.2. Toughness can be regarded as the energy absorbed by the specimen when a certain rotation occurs and can be represented by the area under the load–deflection curve. Residual strength refers to the strength of the specimen when the load decreases to a stable value after surpassing its peak in the load–deflection curve. The calculated results for the relevant parameters are presented in Table 6.

Table 6. Results of toughness and related parameters calculation.

Specimen No.	N_{cu} (N)	Δ_{cu} (mm)	τ_{cu} (MPa)	T_b (N·mm)	σ_b (N·mm^{-2})	τ_c (MPa)
GMPF0-0	2117.6	0.55	4.96	262.095	1.19	0.15
GMPF0.5-15	2259.2	0.619	5.30	308.739	1.25	0.34
GMPF1.0-15	2365.4	0.421	5.54	326.805	1.94	0.31
GMPF1.5-15	2518.8	0.52	5.90	334.398	1.61	0.47
GMPF2.0-15	2748.8	0.641	6.44	456.190	1.78	0.54
GMPF3.0-15	2374.4	0.547	5.57	779.261	3.56	0.90
GMPF1.0-6	2583.6	0.426	6.06	312.085	1.83	0.19
GMPF1.0-9	2352.2	0.421	5.51	264.699	1.57	0.35
GMPF1.0-12	2281.2	0.437	5.35	383.111	2.19	0.39
GMPF1.0-15	2427.2	0.556	5.69	326.805	1.47	0.31
GMPF1.0-19	2257.8	0.378	5.29	300.754	1.99	0.38

Note: N_{cu} represents the maximum bending force (N); Δ_{cu} denotes the corresponding deflection at the maximum bending force (mm); τ_{cu} stands for flexural strength (MPa); T_b represents the absolute toughness value (N·mm); σ_b denotes the flexural toughness coefficient (N·mm^{-2}); and τ_c signifies residual strength (MPa).

The results of the GMPF toughness coefficient calculations are presented in Figure 7, illustrating a noticeable enhancement in the flexural toughness coefficient with fiber incorporation. When the fiber content is 1.0 vol%, the flexural toughness coefficient increases by 63.02% compared to the control group. Similarly, at a fiber content of 3.0 vol%, the flexural toughness coefficient experiences a remarkable increase of 198.35%. This indicates a significant reinforcement effect of polypropylene fibers on the flexural toughness of geopolymer mortar, with the enhancement effect increasing proportionally with the fiber content, mirroring findings similar to those reported by Sukontasukkul et al. [9] in fly ash-based geopolymers. The effect of fiber length on toughness coefficient variations exhibits a fluctuating pattern, with enhancements ranging from 23.52% to 84.03% compared to the GMPF0-0 group as the length increases from 6 mm to 19 mm. The enhancement effect of polypropylene fiber length on flexural toughness is notably significant compared to the control group, while variations within the same group are minor. For instance, compared to the GMPF1.0-12 group, the GMPF1.0-15 group experiences a 48.97% increase in flexural toughness coefficient. Experimental results demonstrate that a fiber length of 12 mm yields the most significant enhancement in flexural toughness.

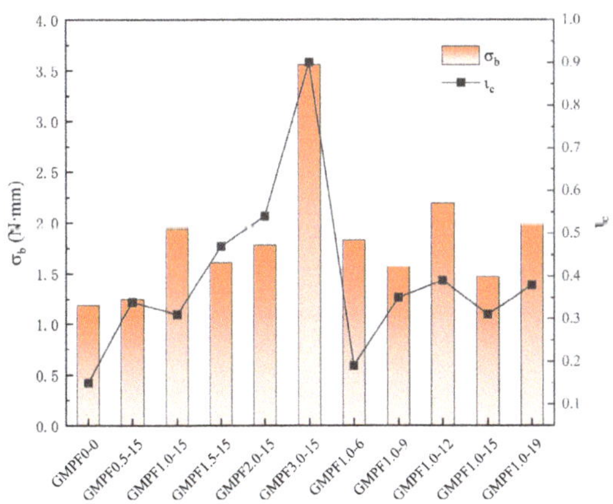

Figure 7. Flexural toughness coefficient and residual strength of GMPF.

Figure 6 illustrates the relationship between residual strength and different fiber contents and lengths. Experimental results indicate an increase in residual strength with higher fiber content. Compared to the control group, residual strength increases by 123.24%, 103.05%, 206.72%, 249.54%, and 484.09% with fiber contents of 0.5 vol%, 1.0 vol%, 1.5 vol%, 2.0 vol%, and 3.0 vol%, respectively. Simultaneously, the influence of fiber length on residual strength grows with increasing length. As fiber length increases from 6 mm to 19 mm, the rate of residual strength improvement ranges from 103.05% to 152.59%. The effect of fiber length on the growth of flexural residual strength appears relatively gradual, attributed to the enhancement of post-peak performance by longer fibers. Longer fibers can provide more interlocking force and energy absorption after the material reaches its peak strength, thereby strengthening the interface bonding between fibers and the matrix, consequently delaying material failure.

In summary, residual strength denotes a specimen's capacity to resist loads after failure, and the inclusion of fibers significantly improves the post-failure stress condition of fly ash-based geopolymers, allowing for a residual flexural strength even after failure. Moreover, considering that fibers in mortar form a robust supporting structure with aggregates, this enhances the overall integrity of the matrix structure, reduces significant deformation of mortar, and mitigates the brittle failure characteristics of fly ash-based geopolymers.

3.4. Microscopic Analysis

3.4.1. SEM Analysis

SEM analysis was conducted on the non-fiber-incorporated group, GMPF0-0, and the fiber-incorporated group, GMPF1.0-15, with the results depicted in Figure 8. From the SEM images of the GMPF0-0 group, it is evident that the compressed GMPF matrix is surrounded by a large amount of amorphous polymer cementitious material. The gel is predominantly irregularly clustered, covering the surface of the matrix, thus densifying the internal structure of the matrix. Additionally, in Figure 8a, unreacted CG particles are observed, existing within the matrix as fillers, reducing micro-defects such as cracks and voids, consequently enhancing the compactness of the structure. Nevertheless, Figures 8b and 8c still reveal the presence of numerous cracks, where larger and wider cracks are primarily due to compressive failure of the matrix, while finer cracks result from significant curing shrinkage of the GMPF matrix itself.

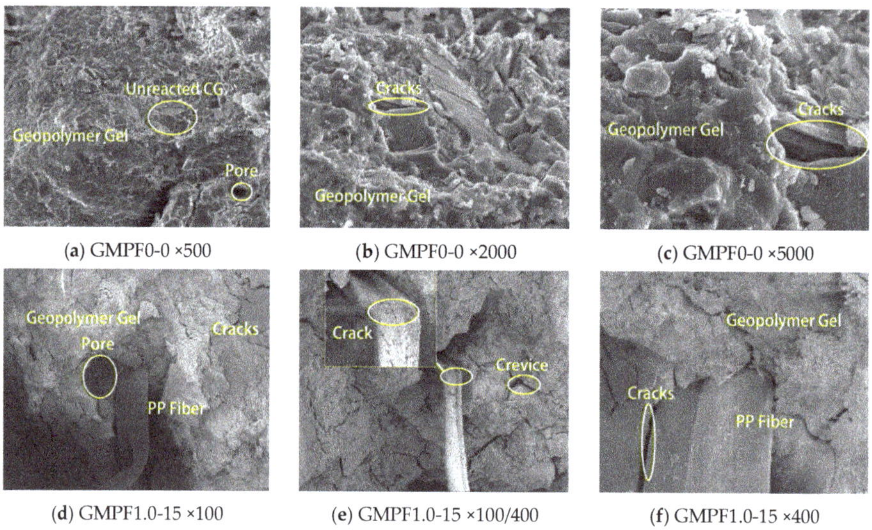

Figure 8. SEM images of GMPF at different magnifications.

Figure 8d–f illustrate the microscopic structure of GMPF fibers with a content of 1.0 vol% and a length of 15 mm. It can be observed that the addition of fibers significantly reduces the occurrence of micro-cracks. The highly elastic PP fibers are distributed within the matrix, optimizing the stress distribution within the matrix, mitigating stress concentration at matrix defects, thus enhancing the compressive strength of the specimens. This is consistent with the trend observed in macroscopic mechanical property changes; when the fiber content is 1.0 vol%, compressive strength increases by 5.31%, and flexural strength increases by 8.2%. However, due to the presence of non-polar chemical bonds on the surface of PP fibers, they exhibit an extremely hydrophobic surface structure. This hydrophobic surface weakens the interaction between PP fibers and the matrix, resulting in the existence of partial cracks and voids between the fiber–matrix interface, leading to a decrease in the density of the matrix structure.

In Figure 8d,f, it can be observed that there is no adherent polymer gel on the fiber surface, and numerous small cracks and voids exist between the fiber and the matrix, weakening the interface bonding ability of the fiber and the matrix, thus diminishing the bridging effect of the fiber. In Figure 8e, it is observed that the connection ends of the fibers exhibit significant tearing and friction marks after loading, with multiple longitudinal cracks appearing on the fiber surface, indicating that polypropylene fibers still possess strong mechanical interlocking capability with the geopolymer matrix. This may be a reason for the slight increase in flexural strength of GMPF specimens but a significant improvement in flexural toughness.

3.4.2. Pore Size and Pore Throat Distribution

The pore size and pore throat distribution of the GMPF specimens were analyzed using the low-field nuclear magnetic resonance test method, and the analysis results are presented below.

Figure 9a illustrates the variation in pore distribution among different groups of GMPF. It can be observed from the figure that the pore distribution curve roughly consists of three peaks, with peak values corresponding to pore sizes of 0.01 μm, 0.5 μm, and 10 μm, respectively. Therefore, following the classification method for pore sizes in cementitious materials, pores ranging from 0 to 0.1 µm are termed micropores, pores ranging from 0.1 to 10 µm are termed mesopores, and pores larger than 10 µm are termed macropores [25]. Additionally, it is noted that peak pores are mainly distributed around the peak near 0.01, followed by the peak near 0.5 µm, while the distribution of pores near 10 µm is minimal. This indicates that the overall pore structure of the GMPF matrix is predominantly composed of micropores, with relatively fewer mesopores and macropores, suggesting a compact overall structure. Compared to the reference group, with the increase in fiber content, the content of micropores in the matrix gradually increases, and the median pore size peak of mesopores shifts to the left, while the median pore size peak corresponding to macropores shows a trend of rightward movement. This indicates that fibers in the matrix can effectively fill the pores of mesopores, converting mesopores with pore sizes ranging from 0 to 10 µm into mesopores with smaller pore sizes ranging from 0.1 to 1 µm and micropores below 0.1 µm. The trend of an expanding pore size of macropores is due to the segmented mesopores merging with other pores in the geopolymer precursor state, transforming mesopores into macropores. The impact of different fiber lengths on pore structure has not varied significantly, with a more significant effect observed on micropores. Generally, longer fibers lead to a higher content of micropores, while changes in mesopores and macropores are minimal, indicating that the internal pore structure of the matrix has not deteriorated.

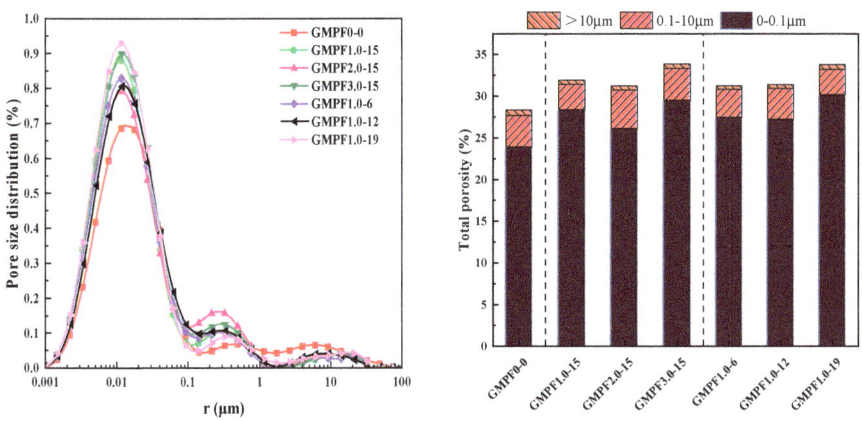

(**a**) Pore size distribution (**b**) Pore throat distribution

Figure 9. Distribution of different GMPF pore types and pore throat sizes.

Table 7 presents the porosity and distribution of harmful pores in GMPF specimens. The definition of harmful pores follows the classification method summarized by Academician Wu Zhongwei, categorized based on the pore size's damaging effect on the mechanical strength of concrete structures, into harmless pores ($0 \leq r < 20$ nm), slightly harmful pores ($20 \leq r < 100$ nm), harmful pores ($100 \leq r < 200$ nm), and multiple harmful holes ($r \geq 200$ nm). The data results show that the porosity increases with the increase in fiber content. When the fiber content increases to 1.0 vol%, 2.0 vol%, and 3.0 vol%, the total porosity of the matrix increases by 11.89%, 9.34%, and 18.64%, respectively. This is because the addition of fibers during specimen preparation introduces some small air bubbles, which are difficult to compact, thus remaining in the matrix. However, considering the combination with mechanical properties, the increase in porosity has not excessively reduced the compressive strength of the GMPF matrix. Its compressive strength still shows improvement compared to the reference group, indicating that the enhancement effect of PP fibers on the mechanical properties of the GMPF matrix outweighs the weakening effect caused by the increase in porosity. On the other hand, due to the hydrophobic nature of polypropylene fiber surfaces, there are gaps between the fibers and the matrix bonding surface, which form pores during geopolymer curing. This phenomenon has been observed in the SEM images of fiber–matrix bonding surfaces in Section 3.4.2. The analysis shows that the longer the fibers, the more pronounced this phenomenon, and more bubbles are introduced when fibers are added. Therefore, when the fiber length is 19 mm, the porosity increases by 18.6% compared to the reference group.

Table 7. Porosity and distribution of harmful pores in GMPF.

Specimen No.	Total Porosity (%)	Harmless Hole (%)	Slightly Harmful Pores (%)	Harmful Hole (%)	Multiple Harmful Holes (%)
GMPF0-0	28.59	15.43	8.49	0.53	4.14
GMPF1.0-15	31.99	20.51	7.86	0.74	2.88
GMPF2.0-15	31.26	18.37	7.72	1.30	3.86
GMPF3.0-15	33.92	20.55	8.94	1.02	3.42
GMPF1.0-6	31.35	19.18	8.27	0.91	2.99
GMPF1.0-12	31.45	18.49	8.71	1.02	3.22
GMPF1.0-19	33.91	29.80	1.63	0.76	1.72

Note: harmless hole ($0 \leq r < 20$ nm), slightly harmful pores ($20 \leq r < 100$ nm), harmful hole ($100 \leq r < 200$ nm), multiple harmful holes ($r \geq 200$ nm).

The calculation of the harmfulness of GMPF pore sizes reveals that pores in the geopolymer matrix primarily appear in the form of harmless and slightly harmful pores, with harmful and multiple harmful holes accounting for only 4.67%. This indicates that the internal structure of the fly ash-based geopolymer matrix is sound, and the polymerization reaction is sufficient, allowing the formed geopolymer gel to effectively fill harmful pores, forming a robust skeletal structure. With the addition of polypropylene fibers, the pore structure of the fiber–geopolymer composite material is significantly improved. When the fiber content increases from 1.0 vol% to 3.0 vol%, the proportion of harmless and slightly harmful pores increases from 23.94% to 28.37%, 26.09%, and 29.45%, respectively, while the proportion of multiple harmful holes decreases from 4.14% to 2.88%, 3.86%, and 3.42%. This is mainly attributed to the filling effect of the fibers in the matrix, converting multiple harmful holes into less harmful, slightly harmful, and harmless pores, optimizing the internal pore structure of the specimens, and increasing the density of the GMPF matrix. Therefore, the fiber-incorporated group exhibits higher mechanical performance. When the fiber content is fixed at 1.0 vol%, varying fiber lengths show a more significant effect on pore structure changes. When the fiber lengths are 6 mm, 12 mm, 15 mm, and 19 mm, the proportion of harmless and slightly harmful pores increases from the original 23.94% to 27.45%, 27.2%, 28.37%, and 31.43%, respectively, while the proportion of multiple harmful holes decreases from 4.14% to 2.99%, 3.22%, 2.88%, and 1.72%, respectively. It is evident that longer fiber lengths result in more harmless and slightly harmful pores and fewer multiple harmful holes, indicating a better internal pore structure of the matrix. Overall, the optimal pore structure is achieved when the polypropylene fiber content is 1 vol% and the length is 19 mm, resulting in the best mechanical strength, consistent with the conclusions in Section 3.2.2.

4. Conclusions

(1) As the PP fiber content increased, more geopolymer paste adhered to the fiber surface, leading to a more pronounced bridging effect in the viscous geopolymer precursor and reducing the fluidity of the GMPF slurry. When the fiber content was less than 1.5 vol%, the fiber length did not significantly affect the flow performance.

(2) The 6 mm polypropylene fibers are more suitable for geopolymers as they disperse more uniformly, forming better interlocking structures and enhancing mechanical properties. The fibers significantly improve the post-failure stress performance and flexural strength of the material, creating a robust support structure within the mortar. This reduces deformation and mitigates brittle failure.

(3) Tears and friction marks were observed at the ends of the fibers where they connect with the matrix after loading, indicating that the fibers effectively suppressed the formation of microcracks. This optimized the internal stress distribution, reducing stress concentration at defects and significantly enhancing the flexural toughness.

(4) LF-NMR results indicated that the incorporation of polypropylene fibers improved the internal pore structure of the geopolymer. As the fiber content increased, the number of small pores increased, and the median pore size of medium and large pores expanded. While fiber length did not alter the overall pore structure, it significantly affected the micropores.

Author Contributions: Conceptualization, J.C.; methodology and validation, A.M.; resources, J.C. and A.M.; data curation, Y.D.; writing—original draft preparation, Y.L.; writing—review and editing, Y.L.; visualization, Y.C.; supervision, H.D. All authors have read and agreed to the published version of the manuscript.

Funding: This research was funded by The Program of Young and Middle-Aged Science and Technology Innovation Leading Talents of Xinjiang Production and Construction Corps, grant number 2019CB023.

Institutional Review Board Statement: Not applicable.

Informed Consent Statement: Not applicable.

Data Availability Statement: The original contributions presented in the study are included in the article, further inquiries can be directed to the corresponding author.

Conflicts of Interest: Author Akelamjiang Maimait was employed by the company Xinjiang Kusha Highway Development Limited Liability Company. The remaining authors declare that the research was conducted in the absence of any commercial or financial relationships that could be construed as a potential conflict of interest.

References

1. Li, Q.; Ren, Z.; Su, X.; Li, P. Bond performance of corroded rebars in sustainable alkali-activated slag-based concrete incorporating steel fibers. *J. Build. Eng.* **2024**, *85*, 108689. [CrossRef]
2. Zhang, D.; Wang, Y.; Zhang, T.; Yang, Q. Engineering and microstructural properties of carbon-fiber-reinforced fly-ash-based geopolymer composites. *J. Build. Eng.* **2023**, *79*, 107883. [CrossRef]
3. Zhang, H.; Chen, W.; He, B.; Zhu, X.; Jiang, Z. Compressive behaviour and microstructure of geopolymer mortar subjected to cryogenic freeze-thaw cycle: Effects of cycles and polypropylene fiber. *Cem. Concr. Compos.* **2024**, *149*, 105505. [CrossRef]
4. Li, H.; Jin, P.; Yan, C.; Zhang, N.; Li, L. Impact of polyethylene fiber and microsilica on fracture properties of high strength high toughness geopolymer concrete (HSHTGC). *Dev. Built Environ.* **2024**, *18*, 100399. [CrossRef]
5. Zhang, X.; Ling, Y.; Wu, Y.; Lai, H.; Wang, Y.; Chen, Z. Comprehensive Analysis of Mechanical, Economic, and Environmental Characteristics of Hybrid PE/PP Fiber-Reinforced Engineered Geopolymer Composites. *Buildings* **2024**, *14*, 1094. [CrossRef]
6. Zhao, X.; Wang, H.; Zhao, X. Sustainable activation of sisal fiber-reinforced slag composites: Mechanical strength and microstructural insights through recycling of alkali-treated wastewaters. *Constr. Build. Mater.* **2024**, *424*, 135971. [CrossRef]
7. Liu, X.; Shi, C.; Yao, Y.; Wei, Y.; Huang, C.; Hong, B. Strengthening Mechanism of Geopolymer Lightweight Cellular Concrete Reinforced with Glass Fibers. *Arab. J. Sci. Eng.* **2024**, 1–16. [CrossRef]
8. Intarabut, D.; Sukontasukkul, P.; Phoo-Ngernkham, T.; Zhang, H.; Yoo, D.-Y.; Limkatanyu, S.; Chindaprasirt, P. Influence of graphene oxide nanoparticles on bond-slip reponses between fiber and geopolymer mortar. *Nanomaterials* **2022**, *12*, 943. [CrossRef]
9. Sukontasukkul, P.; Pongsopha, P.; Chindaprasirt, P.; Songpiriyakij, S. Flexural performance and toughness of hybrid steel and polypropylene fibre reinforced geopolymer. *Constr. Build. Mater.* **2018**, *161*, 37–44. [CrossRef]
10. Kuranlı, Ö.F.; Uysal, M.; Abbas, M.T.; Cosgun, T.; Niş, A.; Aygörmez, Y.; Canpolat, O.; Al-Mashhadani, M.M. Evaluation of slag/fly ash based geopolymer concrete with steel, polypropylene and polyamide fibers. *Constr. Build. Mater.* **2022**, *325*, 126747. [CrossRef]
11. Ayawanna, J.; Poowancum, A. Enhancing flexural strength of metakaolin-based geopolymer reinforced with different types of fibers. *Sustain. Chem. Pharm.* **2024**, *37*, 101439. [CrossRef]
12. Zhang, P.; Ma, L.; Yin, C.L.; Yuan, X.C.; Liu, G.; Zhang, Y.J. Effect of Fiber/hybrid Fiber/sintered Fly Ash on Mechanical Properties of Geopolymer. *Bull. Chin. Ceram. Soc.* **2023**, *59*, 56–62.
13. Mohseni, E. Assessment of Na_2SiO_3 to NaOH ratio impact on the performance of polypropylene fiber-reinforced geopolymer composites. *Constr. Build. Mater.* **2018**, *186*, 904–911. [CrossRef]

14. Al-Mashhadani, M.M.; Canpolat, O.; Aygörmez, Y.; Uysal, M.; Erdem, S. Mechanical and microstructural characterization of fiber reinforced fly ash based geopolymer composites. *Constr. Build. Mater.* **2018**, *167*, 505–513. [CrossRef]
15. Guo, X.; Pan, X. Mechanical properties and mechanisms of fiber reinforced fly ash–steel slag based geopolymer mortar. *Constr. Build. Mater.* **2018**, *179*, 633–641. [CrossRef]
16. *GB/T 18046-2017*; Ground Granulated Blast Furnace Slag Used for Cement, Mortar and Concrete. National Standards of the People's Republic of China: Beijing, China, 2017. Available online: https://www.chinesestandard.net/PDF.aspx/GBT18046-2017 (accessed on 9 June 2024).
17. *ASTM C33-18*; Standard Specification for Concrete Aggregates. ASTM: West Conshohocken, PA, USA, 2023. [CrossRef]
18. *GB/T 2419-2005*; Test Method for Fluidity of Cement Mortar. National Standards of the People's Republic of China: Beijing, China, 2005. Available online: https://www.codeofchina.com/standard/GBT2419-2005.html (accessed on 9 June 2024).
19. Do, V.T.; Nguyen-Tran, H.D.; Chun, D.M. Effect of polypropylene on the mechanical properties and water absorption of carbon-fiber-reinforced-polyamide-6/polypropylene composite. *Compos. Struct.* **2016**, *150*, 240–245. [CrossRef]
20. Xu, S.; Wu, C.; Yue, J.; Xu, Z. Shrinkage and mechanical properties of fibre-reinforced blast furnace slag-steel slag-based geopolymer. *Adv. Civ. Eng.* **2022**, *2022*, 8931401. [CrossRef]
21. Shaikh, F.U.A.; Fairchild, A.; Zammar, R. Comparative strain and deflection hardening behaviour of polyethylene fibre reinforced ambient air and heat cured geopolymer composites. *Constr. Build. Mater.* **2018**, *163*, 890–900. [CrossRef]
22. Guo, R.X.; Guo, J.D.; Yan, F.; Xia, H.Y.; Wang, F.X.; Lin, Z.W.; Ma, Q.M. Investigation on Mechanical Properties and Failure Mechanism of Polypropylene Fiber Reinforced Lightweight Aggregate Concrete. *Bull. Chin. Ceram. Soc.* **2019**, *38*, 1323–1330.
23. Lin, J.X.; Luo, R.H.; Su, J.Y.; Guo, Y.C.; Chen, W.S. Coarse synthetic fibers (PP and POM) as a replacement to steel fibers in UHPC: Tensile behavior, environmental and economic assessment. *Constr. Build. Mater.* **2024**, *412*, 134654. [CrossRef]
24. Alcan, H.G.; Bayrak, B.; Öz, A.; Çelebi, O.; Kaplan, G.; Aydın, A.C. Synergetic effect of fibers on geopolymers: Cost-effective and sustainable perspective. *Constr. Build. Mater.* **2024**, *414*, 135059. [CrossRef]
25. Mindess, S.; Young, F.; Darwin, D. *Concrete*, 2nd ed.; Prentice Hall: Upper Saddle River, NJ, USA, 2003.

materials

MDPI

Article

Simulation Prediction and Experimental Research on Surface Morphology of Ball Head Milling Processing

Youzheng Cui [1,2,3,4,5,*], Xinmiao Li [1], Minli Zheng [2], Haijing Mu [1], Chengxin Liu [1], Dongyang Wang [1], Bingyang Yan [1], Qingwei Li [1], Hui Jiang [1,3], Fengjuan Wang [1,4,5] and Qingming Hu [1,6]

1 School of Mechanical and Electronic Engineering, Qiqihar University, Qiqihar 161006, China; 18646407893@163.com (X.L.); muhaijinghust@163.com (H.M.); liuchengxin1088@163.com (C.L.); 19190706213@163.com (D.W.); 18932293030@163.com (B.Y.); 18814663144@163.com (Q.L.); qzsk_jh@163.com (H.J.); 02742@qqhru.edu.cn (F.W.); 02510@qqhru.edu.cn (Q.H.)
2 Key Laboratory of Advanced Manufacturing and Intelligent Technology, Ministry of Education, Harbin University of Science and Technology, Harbin 150080, China; minli@hrbust.edu.cn
3 QiQiHar Heavy CNC Equipment Co., Ltd., Qiqihar 161005, China
4 The Engineering Technology Research Center for Precision Manufacturing Equipment and Industrial Perception of Heilongjiang Province, Qiqihar 161006, China
5 The Collaborative Innovation Center for Intelligent Manufacturing Equipment Industrialization, Qiqihar 161006, China
6 School of Mechatronics Engineering, Harbin Institute of Technology, Harbin 150001, China
* Correspondence: cuiyouzhenghust@163.com

Abstract: With the aim of improving the machined surface quality of die steel, this paper takes Cr12MoV quenched die steel as the research object and proposes a ball head milling surface morphology prediction model that comprehensively considers influencing factors, including tool vibration, eccentricity, as well as deformation. By setting key parameters, such as line spacing, feed per tooth, cutting depth, and phase difference, the system analyzed the influence of each parameter on the residual height and surface roughness of the machined surface. High-speed milling experiments were conducted, and the surface morphology of the samples was observed and measured under a microscope. The simulation results show good agreement with the experimental data, with errors within 7%~15%, proving the accuracy of the model. This study can provide theoretical support and methodological guidance for surface quality control and processing parameter optimization in complex mold surface machining.

Keywords: surface morphology simulation; ball-end milling; Cr12MoV; Z-map algorithm; tool vibration

1. Introduction

Quenched steel is known for its high hardness, wear resistance, and ease of machining, making it extensively applied in industries, like aerospace, construction machinery, and automotive manufacturing [1,2]. In the automotive industry, quenched die steel is one of the indispensable processing tools. The surface condition after high-speed milling is crucial for the study of surface integrity and is also a key factor in ensuring component performance [3]. Moreover, it exerts a direct influence on the operational lifespan of the mold. Therefore, it is particularly crucial to delve into the morphological characteristics of the processed surface in order to accurately grasp its mechanism of action and make precise predictions of surface morphology. This not only helps to improve the accuracy of mold processing, but also effectively reduces production costs, which has important theoretical value and practical guidance significance.

In recent years, significant progress has been made in the research of three-dimensional surface morphology simulation, which has significantly improved the surface quality and prediction accuracy of complex curved surface part machining. Quinsat et al. [4,5] introduced a 3D surface morphology simulation model that considers tool rotation and translation during machining, yielding more accurate results; their model also allows machining parameters to be adjusted based on the predicted surface roughness. Zhuo, Y. [6] and others have developed a new prediction method that takes into account the vibration and material removal effects in the course of cutting. This method is particularly suitable for modeling and testing the surface morphology of thin-walled parts during peripheral milling. Experiments have shown that it has good accuracy and practicality. Urbikain, G. [7] and others combined geometric modeling and empirical modeling to predict the surface roughness when machining free-form surfaces using a new barrel end mill. They also included the factor of tool runout. By using geometric models based on tool travel paths and mathematical models based on cutting conditions, these models have been validated through experiments on Al7075T6 aluminum alloy and Ti6Al4V titanium alloy, demonstrating high-precision predictive ability. The error of geometric models usually does not exceed 15%, while the error of empirical models does not exceed 10%, among which the feed rate exhibits a notably pronounced effect on the resulting surface roughness. This study provides a reliable tool for efficient and precise machining of complex curved parts. Buj-Corral et al. [8] constructed a numerical model based on the principle of intersection between geometric cutting tools and workpieces, investigating the influence of feed rate and radial depth of cut on the surface topography generated during ball-end milling operations. The results showed that lower surface roughness and shorter processing time can be achieved simultaneously at smaller cutting depths and higher feed rates, which helps to reduce milling and polishing costs and improve processing efficiency and surface quality. Zhang Xiangyu et al. [9] employed an orthogonal experimental design to examine how various process parameters (spindle speed, feed rate, milling depth, milling width) influence the surface roughness of a tin-bismuth alloy. They identified an optimal parameter combination that reduced surface roughness to 0.233 μm, markedly improving the machined surface quality. Wang et al. proposed a method to control surface roughness in five-axis milling through surface morphology analysis. A predictive model was constructed to simulate surface topography based on an elliptical tool trajectory, with a focused examination of how tool runout and workpiece curvature impact the resultant surface characteristics. By solving nonlinear over-determined equations, they optimized the feed rate to ensure a uniform roughness distribution at various positions on the workpiece. This approach improves the uniformity and quality of surface finish while maintaining steady machining efficiency. Wang, L. [10] et al. devised a technique to regulate surface roughness in five-axis milling by analyzing surface features. Researchers worldwide have carried out extensive studies concentrating on modeling and simulating milled surface topography, with the aim of accurately analyzing machined surfaces and effectively controlling their quality. Zhang Wei et al. [11] constructed a model for secondary machining of surface texture using ellipsoidal milling cutters and explored how various milling parameters affect surface texture characteristics. The accuracy of texture simulation was verified through milling experiments. Juan Wei et al. [12] developed a model to predict surface characteristics in ball-end milling using homogeneous coordinate transformation matrices, examining how different parameters influence the resulting surface morphology. Sylvain et al. [13] developed a model for predicting milling surfaces using N-buffer technology, which can take into account local fluctuations in feed rate and is suitable for predicting surface features of free-form surfaces. Gao et al. [14] applied a Z-map algorithm alongside milling experiments to study how cutting impact and tool wear affect surface morphology. They found the simulated data closely matched experimental results, confirming the prediction model's accuracy. This study establishes a theoretical framework to support the optimal selection of machining parameters. Lei Zhang et al. [15] studied the influence and prediction of tool wear on the surface roughness of workpieces. The tool wear area and wear position angle have a significant impact on the surface roughness of workpieces. When the wear area and position angle are controlled within a reasonable range, it actually helps to improve the surface quality of workpieces. Dong et al. developed an improved Z-map algorithm that does not require segmentation of cutting tools for simulating milling surfaces and confirmed its effectiveness and high consistency with simulated surface features through testing [16]. Lazoglu et al. proposed an analytical model for ball-end milling, integrating key machining parameters—including feed rate, flute count, step-over distance, and cutting depth—to predict surface topography and roughness outcomes [17]. Felh ő et al. carried out experimental studies on surface roughness modeling in face milling by systematically adjusting process parameters, including feed rate, cutting speed, and axial depth of cut. A predictive model was established based on response surface methodology, which enabled precise prediction of the surface roughness after milling and clarified the individual effects of each cutting parameter on the surface roughness characteristics [18]. Scholars, such as Yu Yongquan, have explored the surface characteristics of non-circular gears and found that the determining factors for the remaining height are the feed interval and feed rate per turn [19]. Xu Jin et al. [20] carried out a comprehensive investigation on the mechanism underlying surface texture formation during the milling of ultra-high-strength steel AF1410. The study also pointed out how geometric elements and dynamic deformation jointly affect surface roughness in milling operations, and the error range of simulation results was controlled within 10.9%, thus verifying the practicality of the model. Li Yuan et al. explored the principle of surface forming at different positive angles during nanosecond laser milling of aluminum alloys [21]. At the same time, Lu Xiaohong and her team [22] constructed a model to simulate surface morphology. After actual milling experiments, the model demonstrated high accuracy.

Through a comprehensive review of the existing literature, it is found that most of the existing surface topography simulation models are based on idealized assumptions, which often simplify the dynamic behavior of the tool and ignore the influence of important factors, such as tool vibration, eccentricity error, and deformation. Although Cr12MoV die steel is widely used in die and mold processing because of its high hardness and good wear resistance, there is still little research on the evolution of surface morphology, and there is a lack of a three-dimensional surface morphology prediction model that can comprehensively consider the dynamic influence of multiple factors. In order to make up for the above shortcomings, a surface morphology prediction model suitable for Cr12MoV die steel ball end milling was constructed. The proposed model incorporates factors, such as tool vibration, spindle eccentricity, and tool deformation. Based on the cutting edge trajectory modeling and Z-map algorithm, the simulation and prediction of 3D surface topography were implemented using a MATLAB (MathWorks, Natick, MA, USA, Version R2023a) platform. This method provides an effective theoretical basis and reference for optimizing processing parameters and reducing processing costs.

2. Materials and Methods

2.1. Establishing a Mathematical Model for the Cutting Edge in Ball-Head Milling

A precise description of the milling motion is the core requirement for developing a surface morphology simulation model in ball-end milling. For this purpose, it is critical to establish the trajectory equation of the cutting edge of the ball-end mill. Such an equation accurately describes the motion track of each discrete point along the cutting edge during milling. In this work, the center of the ball-end cutter is defined as the origin of the coordinate system, since the ball-end milling cutter contacts the workpiece through its cutting edges. As illustrated in Figure 1, Point Q is any point on the cutting edge. The distance from point Q to the coordinate system origin O is the radius $\boldsymbol{R}$ of the ball head. The included angle between the straight line OQ and axis Z is represented by $\boldsymbol{\theta}$. In the figure, after any point Q of the cutting edge of the tool is projected to the XY plane, the included angle between the projection point Q_1 and point O and axis X is $\boldsymbol{\varphi}$, the Z-axis is consistent with the cutter's axial direction. The cutting edge is divided into n discrete points, and the coordinates of each point Q are expressed as ($\boldsymbol{x}$, $\boldsymbol{y}$, $\boldsymbol{z}$).

$$\begin{bmatrix} x \\ y \\ z \end{bmatrix} = \begin{bmatrix} R\sin\theta\cos(\tan\beta\ln(\cot(\theta/2))) \\ R\sin\theta\sin(\tan\beta\ln(\cot(\theta/2))) \\ -R\cos\theta \end{bmatrix} \tag{1}$$

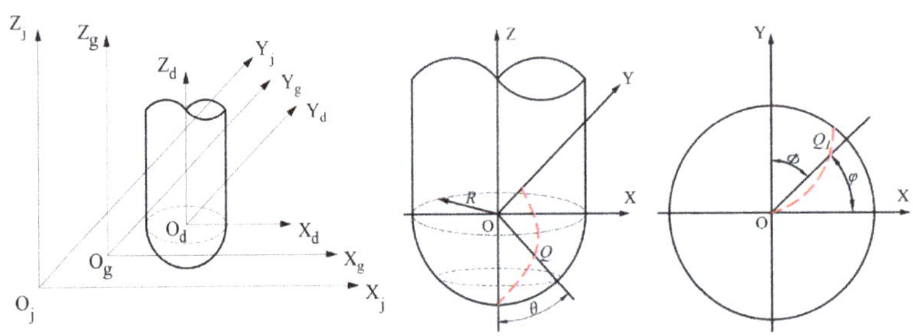

Figure 1. Geometric coordinate model of ball end milling cutter.

In the equation, ($\boldsymbol{x}$, $\boldsymbol{y}$, $\boldsymbol{z}$) represents the coordinates of point Q at any location within the tool coordinate system, $\boldsymbol{\theta}$ denotes the angular position of Q along the cutting edge, and $\boldsymbol{R}$ refers to the radius of the ball-end portion of the ball-end milling cutter.

The angle formed between the line connecting the projection point Q_1 to the origin O and the positive direction of the Y-axis is defined as the phase angle. If the tool has z cutting edges, the phase angle corresponding to the i-th cutting edge can be described by Equation (2):

$$\varphi_i = \varphi_1 + 2\pi(i-1)/z \tag{2}$$

In this study, we define the phase difference ($\Delta\varphi$) as the angular offset between adjacent cutter teeth. This parameter is crucial for modeling the overlapping trajectory and resulting surface morphology during the ball-end milling process. The angular spacing between adjacent cutting edges of a ball-end milling cutter is evenly arranged. By rotating the first cutting edge through an angle $\boldsymbol{\varphi_i}$, the position of the i-th cutting edge can be determined.

Consequently, the coordinates of any point Q located on the i-th cutting edge during the machining process can be formulated as:

$$\begin{bmatrix} x_i \\ y_i \\ z_i \\ 1 \end{bmatrix} = \begin{bmatrix} \cos(\varphi_i - \varphi_1) & -\sin(\varphi_i - \varphi_1) & 0 & 0 \\ \sin(\varphi_i - \varphi_1) & \cos(\varphi_i - \varphi_1) & 0 & 0 \\ 0 & 0 & 1 & 0 \\ 0 & 0 & 0 & 1 \end{bmatrix} \quad (3)$$

In the formula, (x_i, y_i, z_i)—Q is the coordinate of the first cutting edge.

During the milling process, the tool tip does not actually participate in cutting but instead squeezes the workpiece [23]. Therefore, in the actual processing, it is often necessary to determine the appropriate processing angle according to the specific requirements. For the convenience of research, In this paper, the machining inclination angle is divided into a lateral inclination angle α and a forward inclination angle β, as illustrated in Figure 2. Introduce angle transformation matrices as M_1 and M_2.

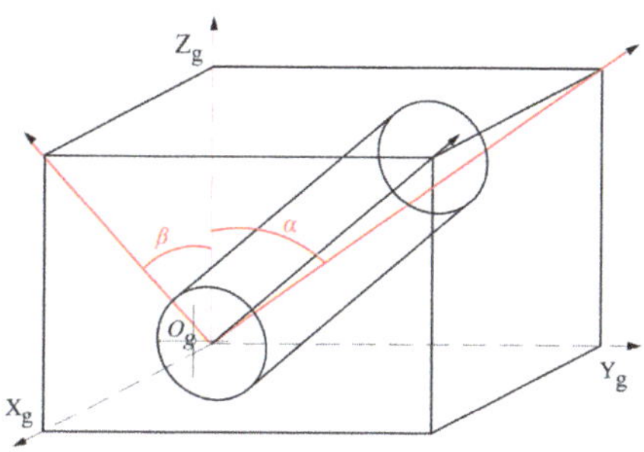

Figure 2. Machining inclination angle of ball nose milling cutter.

By reasonable angle transformation, the transformation matrices of sideslip angle and forward tilt angle can be obtained as follows:

$$M_1 = \begin{bmatrix} 1 & 0 & 0 & 0 \\ 0 & \cos\alpha & -\sin\alpha & 0 \\ 0 & \sin\alpha & \cos\alpha & 0 \\ 0 & 0 & 0 & 1 \end{bmatrix} \quad (4)$$

$$M_2 = \begin{bmatrix} \cos\beta & 0 & -\sin\beta & 0 \\ 0 & 1 & 0 & 0 \\ \sin\beta & 0 & \cos\beta & 0 \\ 0 & 0 & 0 & 1 \end{bmatrix} \quad (5)$$

2.2. Calculation Model for Residual Height of Milling Surface Morphology

Figure 3 illustrates a schematic representation of the residual height in the surface morphology. The arcs *O* and *O1* are the surface morphologies formed between adjacent teeth during the milling process, and point *N* is the vertex of the residual height of the surface morphology. As can be seen, the value of residual height ***MN*** is expressed by the following formula:

$$MN = \frac{8z_0 + \sqrt{64z_0^2 - 16(x_1^2 - 2x_1x_0 + x_0^2 + 4z_0^2 - r^2)}}{8} \quad (6)$$

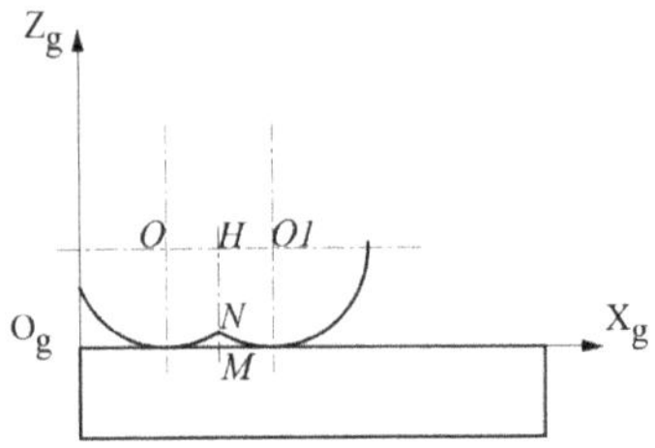

Figure 3. Schematic diagram of residual height of surface morphology.

2.3. Multi Factor Modeling of Ball End Milling Cutter Motion Trajectory

2.3.1. Modeling of the Influence of Tool Vibration on Cutting Edge Motion Trajectory

During milling, tool vibration inevitably alters the cutting trajectory and directly affects the quality of the finished workpiece surface. Therefore, this article assumes that the tool vibrates along the X, Y, and Z directions without interfering with each other, and represents the tool vibration by the displacement of the center of the ball of the ball end milling cutter in the X, Y, and Z directions, in order to construct a tool vibration model. This assumption is based on the following considerations: (1) The stiffness distribution characteristics of the tool structure in the high-speed milling process often lead to the axial vibration to dominate, while the mutual interference effect is relatively weak; (2) This paper focuses on the influence of vibration on the residual height of the surface; (3) Available literature indicates (for example, Sun et al. [24]) that under normal machining parameters, the mutual interference between XYZ vibrations has a limited influence on the surface prediction error. The structural design and dynamic stiffness distribution of high speed milling tools often lead to dominant vibrations in a single direction at any given time, particularly under steady cutting conditions. Therefore, in order to balance the complexity and computational efficiency of the model, the interference of lateral vibration is simplified in this study.

The vibration of the cutting tool is similar to that of a spring damping system, and the dynamic model of the system can be expressed by Equation (7) [24].

$$\begin{array}{l} \frac{d^2x}{dt^2} + 2\xi_x w_x \frac{dx}{dt} + w_x^2 x(t) = \frac{w_x^2}{k_x} F_x(t) \\ \frac{d^2y}{dt^2} + 2\xi_y w_y \frac{dy}{dt} + w_y^2 y(t) = \frac{w_y^2}{k_y} F_y(t) \\ \frac{d^2z}{dt^2} + 2\xi_z w_z \frac{dz}{dt} + w_z^2 z(t) = \frac{w_z^2}{k_z} F_z(t) \end{array} \tag{7}$$

In this formula, $\boldsymbol{F_x(t)}$ represents the cutting force on the X-axis; $\boldsymbol{F_y(t)}$ represents the cutting force on the Y-axis; $\boldsymbol{F_z(t)}$ represents the cutting force on the Z-axis; $\boldsymbol{\xi}$ is the damping coefficient of the vibration system; $\boldsymbol{\omega}$ is the natural frequency of the vibration system under dynamic conditions; $\boldsymbol{k}$ is the modal stiffness of the vibration system under dynamic conditions.

By solving the above vibration equation, the coordinate transformation matrix due to tool vibration can be obtained as:

$$M_3 = \begin{bmatrix} 1 & 0 & 0 & -x(t) \\ 0 & 1 & 0 & -y(t) \\ 0 & 0 & 1 & -z(t) \\ 0 & 0 & 0 & 1 \end{bmatrix} \tag{8}$$

2.3.2. Modeling of the Influence of Tool Eccentricity on Cutting Edge Motion Trajectory

In order to enhance the accuracy of the simulation, potential misalignment between the tool axis and the spindle axis must be considered in the model (i.e., the eccentricity introduced during tool mounting; see Figure 4).

Within the coordinate frame of the workpiece, let e be the radius of the eccentric circle caused by spindle rotation, and let μ be the eccentric angle between the tool and spindle axes. Then, the correction matrix for the tool's eccentric cutting-edge motion can be expressed as:

$$M_4 = \begin{bmatrix} 1 & 0 & 0 & -\mathrm{e}\cos\mu \\ 0 & 1 & 0 & -\mathrm{e}\sin\mu \\ 0 & 0 & 1 & 0 \\ 0 & 0 & 0 & 1 \end{bmatrix} \tag{9}$$

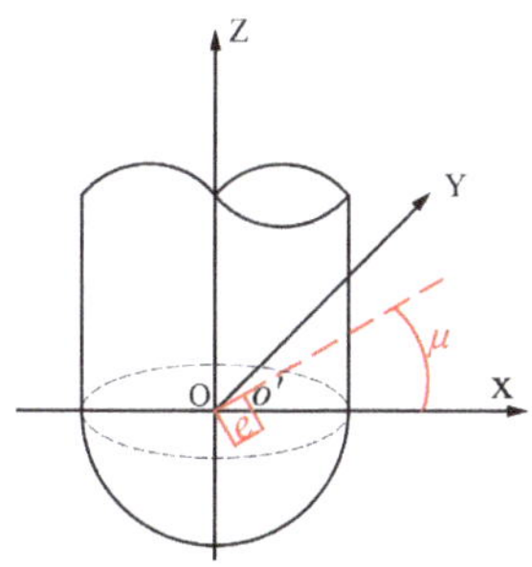

Figure 4. Schematic diagram of eccentricity of ball nose milling cutter.

2.3.3. Modeling of the Influence of Tool Deformation on Cutting Edge Motion Trajectory

In high-speed milling, the tool is subject to vibration and deformation under cutting forces. Let $x(z_j,t)$ and $y(z_j,t)$ denote the tool's deformation in the x and y directions, respectively. The deformation of the ball-end mill in the x-direction is given by [25]:

$$x(z_j,t) = \frac{F_x}{6EI}\left[(r_x(\theta)-z)^3 - (L-Z)^3 + 3(L-Z)^2(L-r_x(\theta))\right] \tag{10}$$

In the formula, I denotes the moment of inertia acting on the ball-end milling cutter; E represents the elastic modulus of the tool material; L denotes the length of the ball-end milling cutter; F_x representing the component of cutting force in the x-direction; $r_x(\theta)$ represents the center position of the instantaneous cutting edge along the x-axis.

In the direction of the z-axis, the surface plastic deformation of the workpiece together with its vibration are the key contributing factors. The plastic strain attains its peak value at the workpiece surface and exhibits a negative exponential distribution. In the direction perpendicular to the workpiece surface, this value rapidly decreases. The plastic strain [26] can be obtained as follows:

$$\begin{gathered} \varepsilon(z) = J_1 e^{-J_2 z} \\ J_1 = \frac{\cos(\alpha_0)}{\sqrt{3}\sin\varphi\cos(\varphi-\alpha_0)} \\ J_2 = -\frac{1}{(n+1)d}\ln\left[1 - \frac{(n+1)\left(F_x\cos\varphi - F_y\sin\varphi\left(\sqrt{3}\sin\varphi\right)^{n+1}\cos^n(\varphi-\alpha_0)\right)}{Ka_e a_p\cos^n\alpha_0}\right] \end{gathered} \tag{11}$$

In the formula, α_0 represents the rake angle of the tool; n represents the hardening index; K represents the yield strength; d indicates the depth of deformation; a_e represents the line spacing, and a_p denotes the cutting depth.

The plastic deformation can be obtained from Equations (12) and (13):

$$\delta = \int_0^d J_1 e^{-J_2 z} dx \tag{12}$$

$$\delta_d = -\frac{J_1}{J_2} e^{-J_2 d} \tag{13}$$

Therefore, the matrix used to correct the influence of tool shape changes on tool paths can be expressed as:

$$M_5 = \begin{bmatrix} 1 & 0 & 0 & u(z_j, t) + x(t) \\ 0 & 1 & 0 & v(z_j, t) + y(t) \\ 0 & 0 & 1 & \delta_d + z(t) \\ 0 & 0 & 0 & 1 \end{bmatrix} \tag{14}$$

After establishing the geometric model of the spherical cutting edge and calculating the transformation matrices, the motion path equation for any point on the tool's cutting edge can be described by the following formula:

$$\begin{bmatrix} x\prime \\ y\prime \\ z\prime \\ 1 \end{bmatrix} = M_5 M_4 M_3 M_2 M_1 \begin{bmatrix} x \\ y \\ z \\ 1 \end{bmatrix} \tag{15}$$

2.4. Modeling Method for Surface Morphology

Given that the surfaces of most milled workpieces belong to free-form surfaces, describing free-form surfaces can be cumbersome. In mechanical manufacturing, there are three primary mathematical representations for free-form surfaces: Bézier, Coons, and NURBS for free-form surfaces: Bezier, Coons, and NURBS (Non-Uniform Rational B-Spline) surfaces. B-spline surfaces have many advantages, especially their ability to locally adjust when depicting complex shape related parts, making NURBS surfaces an ideal choice. Therefore, this article decides to adopt NURBS surface reconstruction based on flexibility to achieve discrete point fitting and obtain surface morphology. The specific steps of this process are shown in Figure 5.

In order to realize the accurate reconstruction of the machining surface, this paper adopts the non-uniform rational B-spline (Non-Uniform Rational B-Splines, NURBS) surface fitting method and reconstructs the workpiece surface morphology by combining the tool trajectory data obtained from simulation. The specific implementation process mainly includes the following four stages: (1) Parameter initialization. Set the basic cutting parameters including feed rate, number of teeth and cutting time, establish the spatial domain of the machining area, and initialize the height matrix H to record the residual height value corresponding to the tool trajectory at each position. (2) Perform trajectory discretization and point cloud generation. By traversing the feed path and time steps, calculate the spatial coordinates of the tool at each moment; determine whether the current coordinate (x, y) falls within the predefined spatial domain. If the condition is met, locate the corresponding position in the height matrix H and compare the calculated height value z with the original value at that position. If the z value is smaller (indicating material removal), update the matrix H and save the point to the point cloud dataset. (3) Perform

NURBS fitting and error correction on the generated point cloud. Input the point cloud data into the NURBS surface reconstruction model and calculate the distance deviation s between each sample point and the initial control surface. If $s < 0$, it indicates that the point is below the theoretical surface, and it needs to be expanded outward along the normal; if $s > 0$, iterate in the direction of the control points until the error meets the convergence criteria. (4) When all the sampling points converge, the NURBS reconstruction process of surface processing simulation is completed, and the final three-dimensional surface morphology simulation results are obtained.

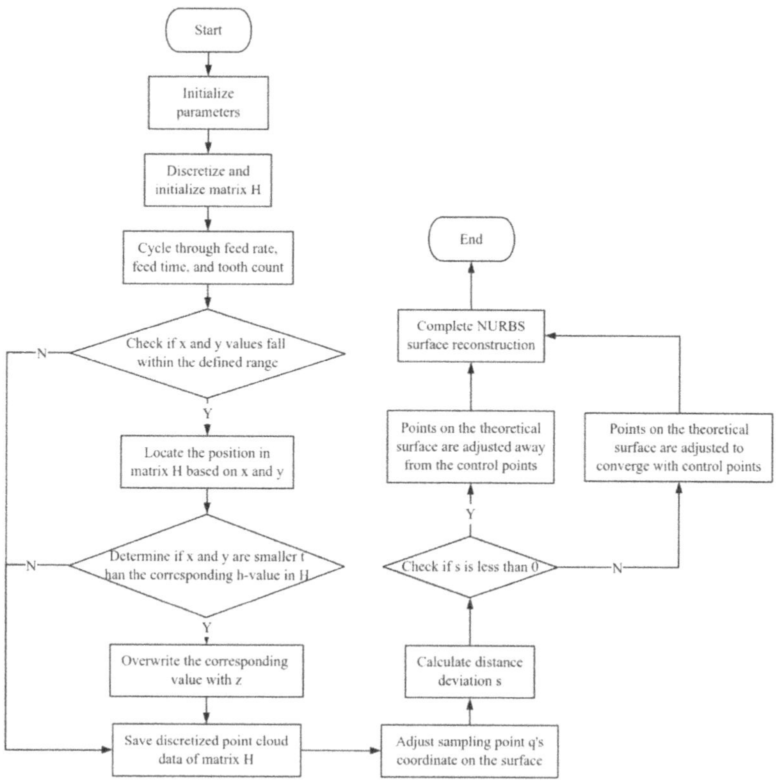

Figure 5. Matlab simulation flowchart of surface morphology.

Compared with traditional Bezier surface, NURBS fitting has higher local control and modeling flexibility, which can better adapt to the processing simulation requirements of complex surface structure.

2.5. Materials and Experiments

The machine tool used in this experiment is the DMU 60mono BLOCK five axis (DMG MORI company, Pfronten, Bavaria, Germany) high-speed vertical machining center, with a spindle speed of up to 10,000 r/min, as shown in Figure 6. The cutting tool used is a SH300-H solid carbide ball-end mill (Xiamen Jinlu, Xiamen, China). A white light interferometer (Figure 7) was used for surface measurement, and the workpiece material was Cr12MoV quenched steel (Daye Special Steel Co., Ltd., Daye, Hubei, China) (70 HRC). See Table 1 for main material parameters.

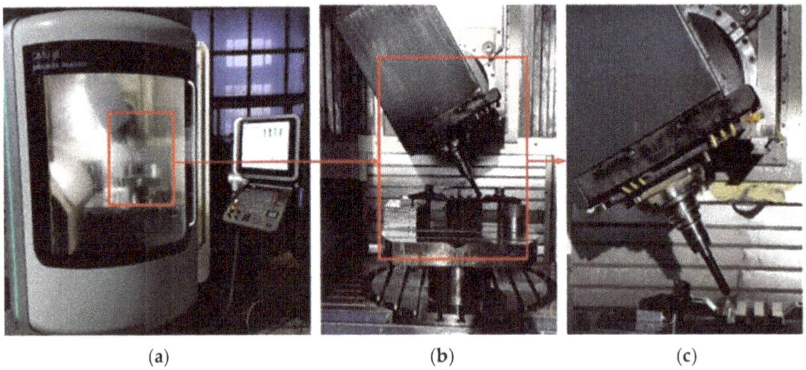

Figure 6. Numerical control machine tool used in the experiment. (**a**) External view of the five-axis CNC milling machine. (**b**) Processing state with 45 ° inclination of main shaft. (**c**) Local enlarged drawing of surface contact.

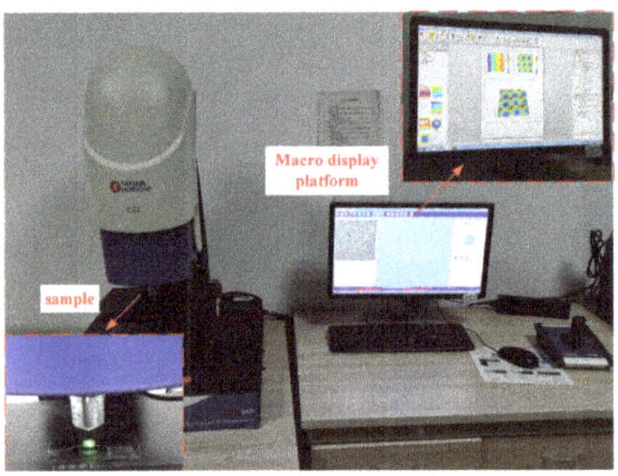

Figure 7. White light interferometer.

Table 1. Chemical composition of Cr12MoV [27].

Element	C	Cr	Mo	V	Mn
content	1.45–1.70	11.00–12.50	0.4–0.6	0.15~0.30	≤0.40
element	Si	Ni	Cu	S	P
content	≤0.40	≤0.25	≤0.30	≤0.030	≤0.030

3. Simulation Analysis of the Influence of Processing Parameters on Milling Surface Morphology

In order to study the influence of different process parameters on milling surface morphology, single-factor simulation and experimental research are carried out in this section. In order to ensure the scientific and engineering relevance of the experimental design, based on the available literature, the milling parameters selected in this study under the same processing conditions are: line pitch (a_e: 0.4–0.7 mm), feed per tooth (f_z: 0.4–0.8 mm/z), and cutting depth (a_p: 0.3–0.6 mm), as shown in Table 2, ensuring the representativeness of the experimental results when milling Cr12MoV die steel at high speed.

Table 2. Single-factor simulation and experimental scheme of different milling parameters.

Serial Number	Line Spacing a_e (mm)	Feed per Tooth f_z (mm/z)	Cutting Depth a_p (mm)	Spindle Speed n (r/min)
a	0.4	0.4	0.3	
b	0.5	0.5	0.4	
c	0.6	0.6	0.5	10,000
d	0.7	0.7	0.6	
e	-	0.8	-	

3.1. Simulation Study on the Influence of Line Spacing on Surface Morphology

In traditional milling processes, due to the large line spacing, machining residues are mainly line spacing residues. When milling mold steel with ball end milling cutters, different line spacing can obtain different surface morphologies. Therefore, this study first analyzed the effect of line spacing a_e on the surface morphology in simulation. To further explore the influence mechanism and variation trend of line spacing a_e on surface morphology, a single factor simulation design scheme for line spacing was carried out as shown in Table 2. The tool was fed unidirectionally with a fixed inclination angle of 30°, spindle speed n = 1000 (r/min), line spacing a_e: 0.4~0.7 mm, feed per tooth f_z = 0.4 mm/z, cutting depth a_p = 0.3 mm, Figure 8a–d illustrates the simulated surface morphologies under these conditions.

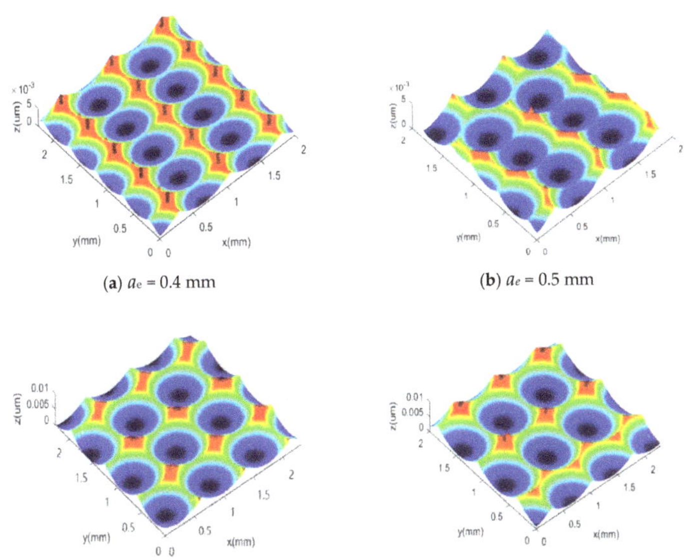

(a) a_e = 0.4 mm (b) a_e = 0.5 mm (c) a_e = 0.6 mm (d) a_e = 0.7 mm

Figure 8. The effect of line spacing on surface morphology (red indicates peaks, and blue indicates valleys).

From the simulation results of Figure 8a–d, it can be seen that when other cutting parameters remain constant and only the milling line spacing a_e changes, the surface morphology is periodically and uniformly distributed along the line spacing direction and feed direction. However, the change in line spacing a_e only affects the surface morphology in the line spacing direction. With the continuous increase of line spacing a_e, the surface morphology undergoes significant changes in the line spacing direction. The length dimension of the micro unit in the line spacing direction gradually increases with the continuous increase of line spacing a_e, and the frequency of residual height in the line spacing decreases with the increase of line spacing a_e. The value of residual height gradually increases, while the length in the feed direction remains basically unchanged.

3.2. Simulation Study on the Influence of Feed per Tooth on Surface Morphology

According to the simulation analysis, it is evident that variations in feed per tooth lead to distinct differences in the resulting surface morphology. During the ball-end milling process, the feed per tooth f_z is a key factor influencing the surface characteristics. To explore how the feed per tooth f_z affects surface formation, a single-factor simulation was performed (see Table 2), line spacing a_e: 0.4 mm, feed per tooth f_z = 0.4~0.8 mm/z, cutting depth a_p = 0.3 mm, Figure 9a–e show the simulated surface morphologies under these feed rates.

The simulation results of surface morphology shown in Figure 9a–e indicate that the surface presents a periodic and uniform distribution pattern in both the feed direction and the line spacing direction. Nevertheless, there are slight differences in the variation law. In contrast to the influence of line spacing on surface contour, the feed per tooth f_z primarily influences the surface profile along the feed direction. With the increase of feed per tooth f_z, the length of the microstructure along the feed direction grows continuously, the frequency of residual height decreases, and the residual height value gradually rises.

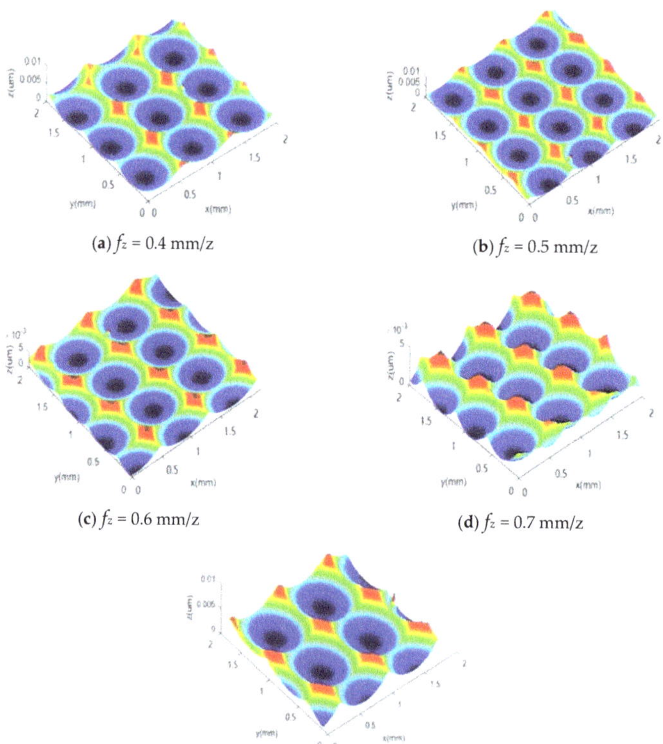

(**a**) f_z = 0.4 mm/z (**b**) f_z = 0.5 mm/z (**c**) f_z = 0.6 mm/z (**d**) f_z = 0.7 mm/z (**e**) f_z = 0.8 mm/z

Figure 9. Effect of changing the feed per tooth on the surface topography (red indicates peaks, and blue indicates valleys).

3.3. Simulation Study on the Influence of Cutting Depth on Surface Morphology

Variations in cutting depth during the milling process have a significant impact on the surface morphology. To investigate how milling depth influences surface characteristics and their changing patterns, this study carried out a single-factor simulation scheme focusing on the variation of cutting depth a_p. The specific data can be observed in Table 2, with line spacing a_e: 0.4 mm, feed per tooth f_z = 0.6 mm/z, and cutting depth a_p = 0.3~0.6 mm. Figure 10a–d shows the simulation results of the workpiece surface morphology.

As shown in the simulation results in Figure 10, it is evident that the length of microstructural units on the machined surface slightly increases with the rise of cutting depth a_p, although this increase is limited. The reason for this phenomenon is that a higher cutting depth a_p causes an enlargement of the milling radius of the ball-end mill engaged in the workpiece, thereby contributing to the growth of the microstructure length. However, since the radius of the cutter's ball end is significantly larger than the cutting depth a_p, the overall increase in microstructure length remains relatively small. Additionally, as the cutting depth a_p continues to rise, the residual height generated by the tool path interval and feed direction also increases correspondingly, resulting in a continuous growth of surface roughness values.

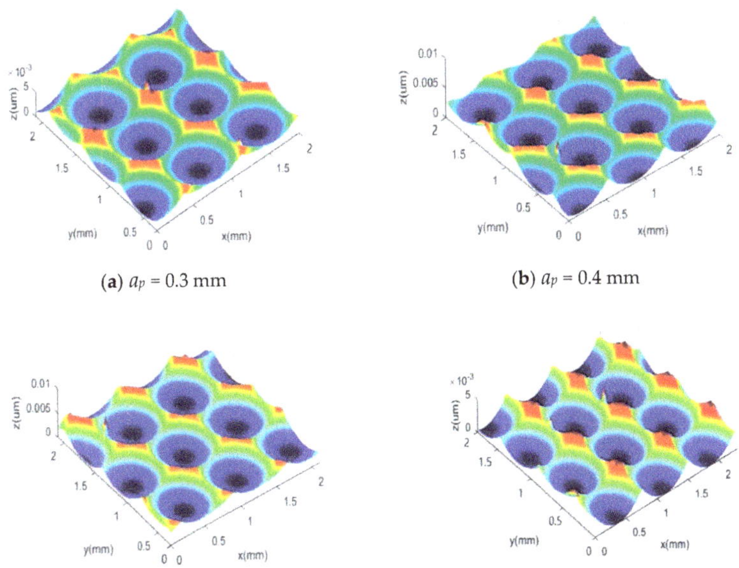

(**a**) a_p = 0.3 mm (**b**) a_p = 0.4 mm

(**c**) a_p = 0.5 mm (**d**) a_p = 0.6 mm

Figure 10. The effect of cutting depth on surface morphology (red indicates peaks, and blue indicates valleys).

3.4. Simulation Study on the Influence of Phase Difference on Surface Morphology

During high-speed ball-end milling, the phase difference $\Delta\varphi$ between adjacent cutting edges plays a crucial role in determining the formation of the workpiece surface microstructure. Traditional simulation models often overlook the relative angular relationship between cutting teeth. In this work, a surface morphology prediction model is established that accounts for the overlapping offsets of adjacent cutter trajectories, which significantly enhances simulation accuracy. This model is used to explore the effects of different initial phase angles φ_0 on cutting trajectory interference behavior and surface residual height (Figure 11).

The figure shows the simulated morphology of ball head milling when $\Delta\varphi$ is randomly set to 0°, 30°, 60°, and 90°. It can be observed that (1) when the phase difference is a certain value, the surface texture direction is always the same, and the surface morphology is arranged in a regular pattern; (2) when the phase difference is random (i.e., the cutting edges are out of phase by a random angle), the surface morphology and texture arrangement become chaotic, with no regular pattern, and no regular texture arrangement is formed; (3) when the phase difference changes from 0° to 90°, the surface morphology gradually changes from a quadrilateral to a hexagon, and the direction angle of the surface texture gradually increases with the increase of the phase difference; (4) when the phase difference is 0°, the surface morphology is a regular quadrilateral morphology, and when the phase difference is 90°, the surface morphology is a regular hexagonal pattern.

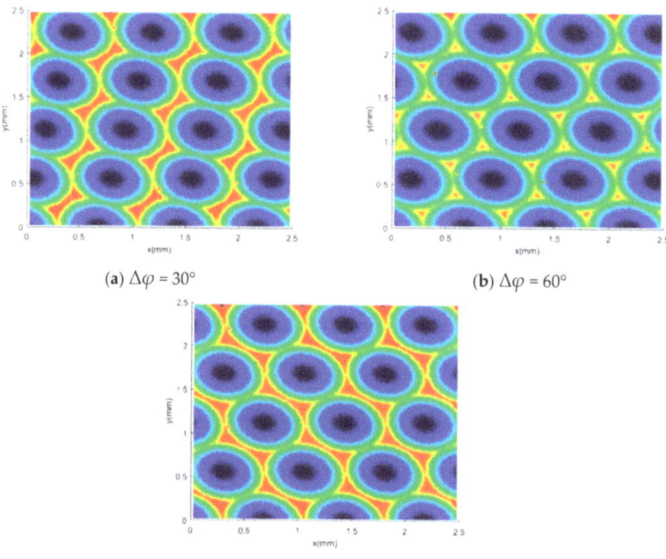

(**a**) $\Delta\varphi$ = 30°

(**b**) $\Delta\varphi$ = 60°

(**c**) $\Delta\varphi$ = 90°

Figure 11. The effect of phase difference $\Delta\varphi$ on surface morphology (red indicates peaks, and blue indicates valleys).

4. Experimental Study on Surface Morphology of High-Speed Ball Head Milling Processing

4.1. Experimental Study on the Influence of Line Spacing on Surface Morphology

In order to validate the accuracy of the simulation results regarding the effect of line spacing on surface morphology, this paper designed a single factor experiment on milling line spacing. Firstly, the samples were machined under the specified conditions, and then the surface morphologies of different experimental samples with milling line spacing were observed under a microscope, as illustrated in Figure 12a–d.

Through the comparison and analysis of the experimental data and the simulation results, it can be concluded that the simulated milling surface morphology is relatively consistent with the experimental findings. Figure 13 presents the variation trend of the three-dimensional surface roughness by comparing the experimental measurements with the simulated values.

As illustrated in Figure 13, the simulated surface roughness values of the workpiece show a high degree of consistency with the experimental measurements. With the continuous increase in line spacing a_e, the surface roughness values exhibit a gradual upward trend. This phenomenon is mainly attributed to the fact that a larger line spacing expands the distance between adjacent tool paths, resulting in more uncut material remaining on the workpiece surface. Consequently, the residual height in the line spacing direction increases, causing the overall three-dimensional surface roughness to rise. The deviation between the simulation results and experimental data is within the range of 7% to 15%, and the simulated values are generally smaller than the corresponding experimental values.

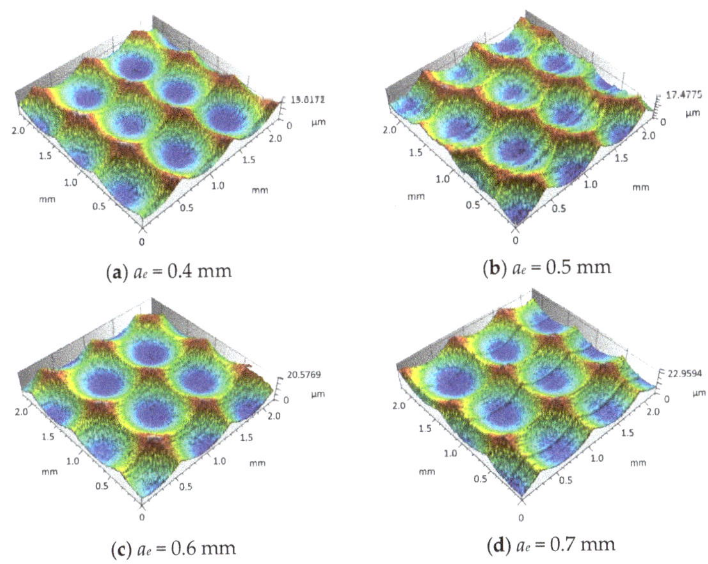

Figure 12. Experimental results of the influence of line spacing on surface morphology.

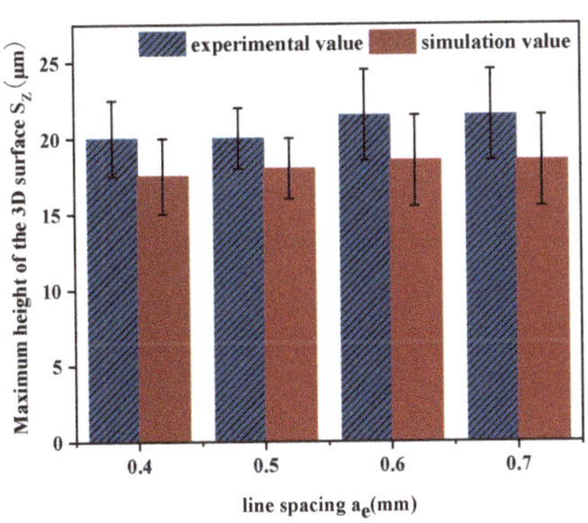

Figure 13. Comparison of surface roughness between different a_e experiments and simulations.

4.2. *Effect of Feed per Tooth on Surface Morphology*

Based on simulation experiments, it is known that the difference in feed per tooth will result in different surface morphologies. To validate the simulation results for feed per tooth, we conducted a single-factor experiment varying the feed per tooth on surface morphology, this paper designed a single factor experiment on feed per tooth. Firstly, the samples were processed for morphology, and then the surface morphology of different experimental samples was observed microscopically to obtain the measured surface morphology of samples with different feed per tooth rates, as shown in Figure 14a–e.

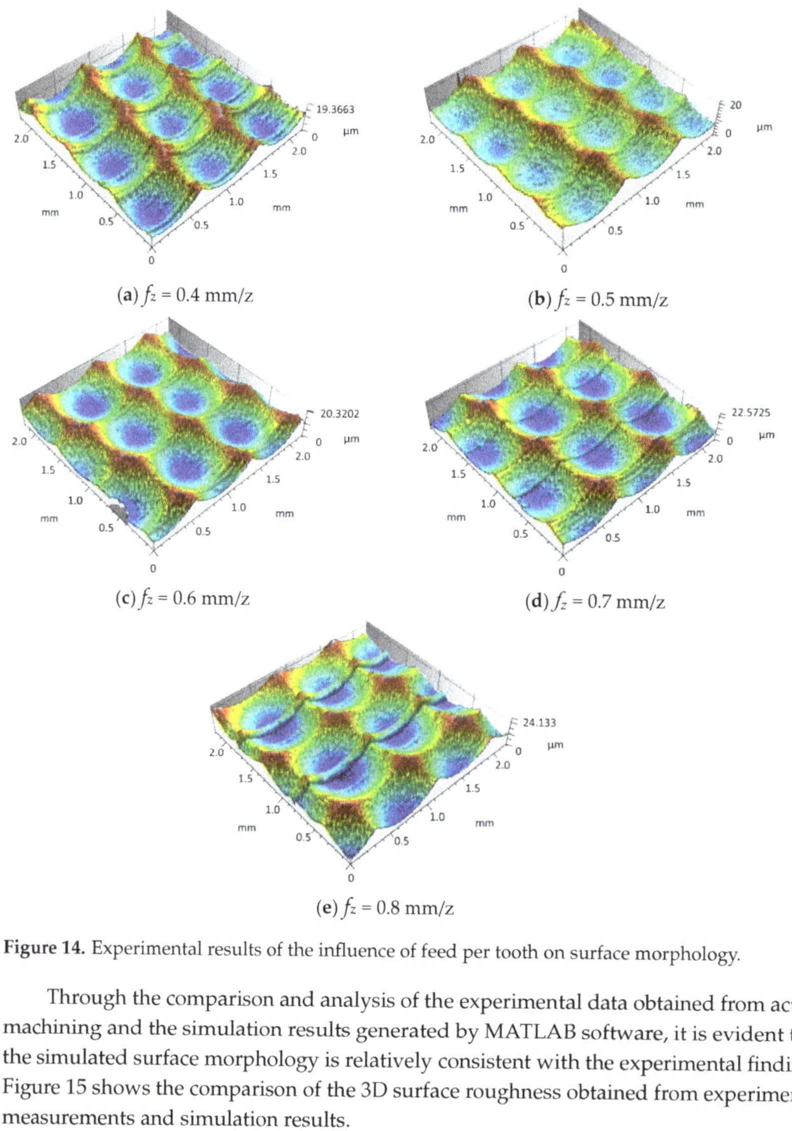

(**a**) f_z = 0.4 mm/z (**b**) f_z = 0.5 mm/z

(**c**) f_z = 0.6 mm/z (**d**) f_z = 0.7 mm/z

(**e**) f_z = 0.8 mm/z

Figure 14. Experimental results of the influence of feed per tooth on surface morphology.

Through the comparison and analysis of the experimental data obtained from actual machining and the simulation results generated by MATLAB software, it is evident that the simulated surface morphology is relatively consistent with the experimental findings. Figure 15 shows the comparison of the 3D surface roughness obtained from experimental measurements and simulation results.

As illustrated in Figure 15, the comparison between the experimental results of surface roughness *Sz* and the simulated data shows that both exhibit a consistent variation pattern. In particular, the surface roughness increases progressively as the feed per tooth fz rises. This behavior is mainly caused by the fact that a larger feed per tooth fz produces a higher residual height along the feed direction, thereby amplifying the surface roughness. However, there remains an error between the experimental and simulated results, ranging from approximately 7% to 15%, primarily due to the increase in cutting force generated by a higher feed per tooth fz during machining. In actual machining, surface morphology with varying depths and sizes is further formed due to factors such as tool deformation and cutting vibration. However, in the simulation cutting process, it is a relatively stable and ideal environment, and the resulting surface morphology is also relatively stable and regular, resulting in certain errors be-tween simulation values and experimental values.

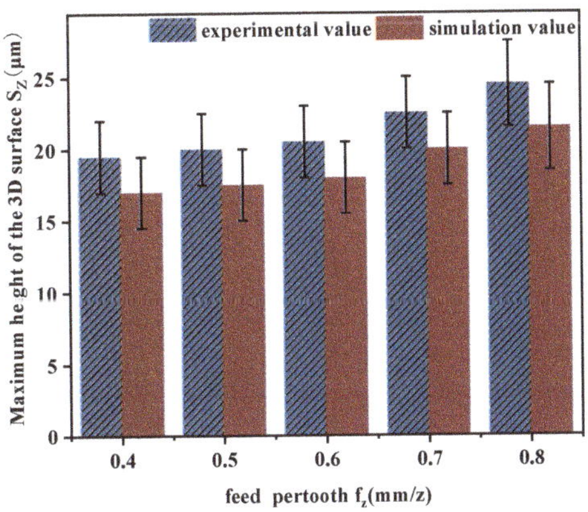

Figure 15. Comparison of surface roughness between fz experiment and simulation.

4.3. The Influence of Cutting Depth on Surface Morphology

To validate the accuracy of the simulation results regarding the effect of cutting depth on surface morphology, this paper designed a single factor experiment on cutting depth. Firstly, the samples were processed for morphology, and then the surface morphology of different experimental samples was observed microscopically. As shown in Figure 16a–d, the measured surface morphology corresponds to samples processed with different cutting depths.

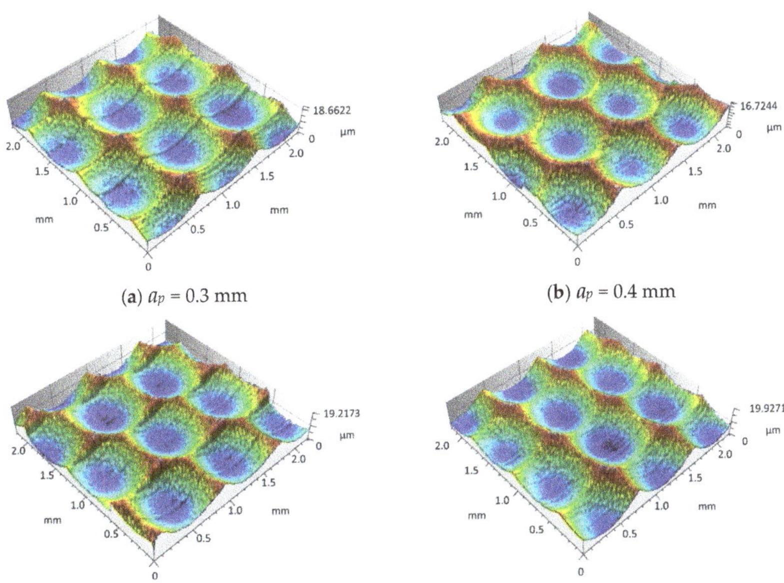

(**a**) a_p = 0.3 mm (**b**) a_p = 0.4 mm (**c**) a_p = 0.5 mm (**d**) a_p = 0.6 mm

Figure 16. Experimental results of the influence of cutting depth on surface morphology.

Through the comparative analysis of the experimental results obtained from actual machining and the simulation results generated by MATLAB software, it is observed that the simulated surface morphology is relatively consistent with the experimental findings. The corresponding comparison of the experimental and simulated values of the three-dimensional surface roughness is illustrated in Figure 17.

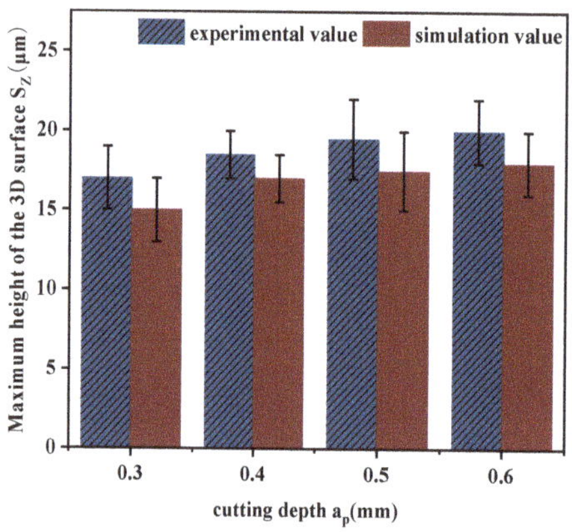

Figure 17. Comparison of surface roughness between a_p experiments and simulations.

Figure 17 presents the comparison of the measured and simulated values of the three-dimensional surface roughness Sz corresponding to different cutting depths. It is observed that with the increase of cutting depth a_p, both experimental and simulated surface roughness values rise accordingly. This is mainly due to the fact that a larger cutting depth results in a higher radial cutting force, which strengthens the tool's motion and vibration, thus causing the surface roughness to increase. Consequently, there is a certain deviation between the simulation and experimental results, ranging from 7% to 15%.

Haining Gao et al. [14] evaluated quenched Cr12MoV steel and found that the maximum deviation between simulated and experimental roughness reached 21.9%, while the maximum deviation in this study was 15%, which is within a controllable range and has good reference value. The main reason for the differences lies in the different experimental equipment used. Three axis CNC machine tools were used, while a five axis high-speed vertical machining center was used in this study. The spindle stiffness of the machine tool was better, the cutting speed was faster, and the maximum spindle speed could reach 18,000 r/min. The vibration was smaller, and the milling process was more stable and conducive to forming a regular surface morphology. The influence of tool wear was not considered (in order to reduce the impact of tool wear on this study, new blades were replaced after each milling). Therefore, the roughness deviation in this study is relatively small.

4.4. The Influence of Phase Angle on Surface Morphology

To further verify the actual effect of phase angle on the surface morphology of ball head milling, this paper designed high-speed milling experiments under different tooth to tooth phase difference conditions. Selecting random angles of 0°, 30°, 60°, and 90° with $\Delta\varphi$ as the research objects, ensuring that other cutting parameters (feed rate, line spacing, cutting depth, etc.) remain consistent.

According to the high-speed ball head milling experiments shown in Figure 18a–e, By comparing and analyzing the experimental data with the simulation outcomes, it is evident that the simulated surface morphology in the milling process is relatively consistent with the experimental observations. From Figure 18, it can be seen that when the phase difference remains constant, the surface texture direction is always the same, and the surface morphology is arranged in a regular pattern. When the phase difference randomly changes, the arrangement of morphology and texture becomes chaotic. Due to the fact that the number of cutting edges of the ball end milling cutter is 2, the morphology gradually changes from a quadrilateral to a hexagon when the phase difference is from 0~90°, and the texture direction gradually increases. When the phase difference is 0, the surface morphology is a regular quadrilateral morphology, and when the phase difference is 90°, the surface morphology is a regular hexagonal morphology.

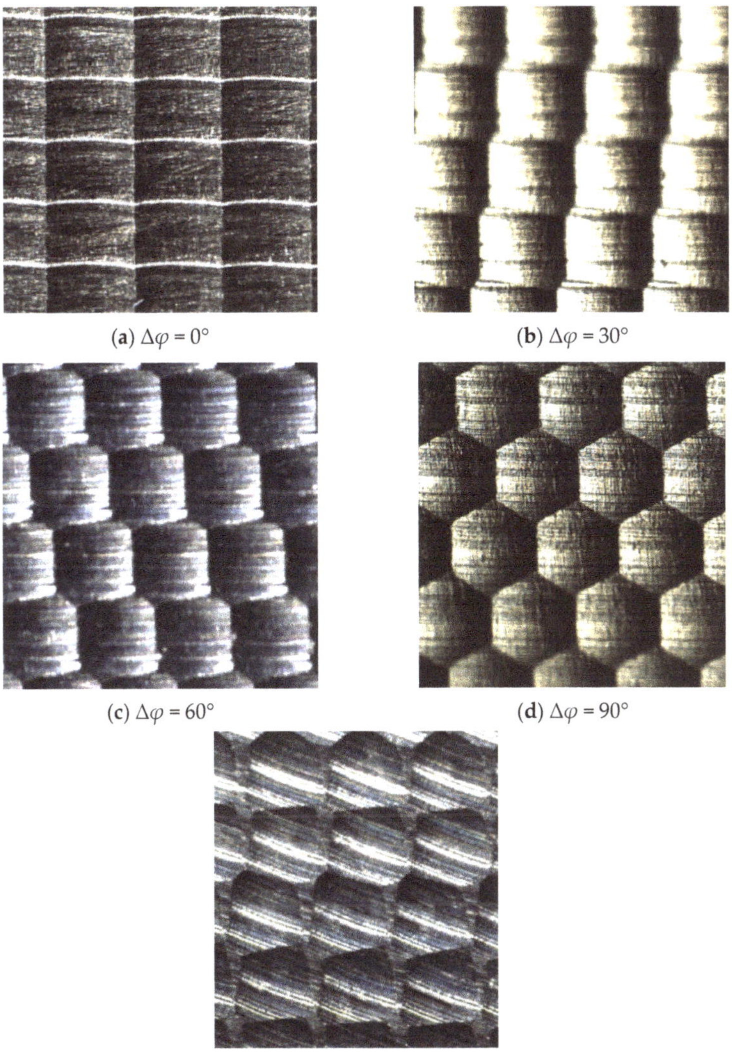

(**a**) $\Delta\varphi = 0°$ (**b**) $\Delta\varphi = 30°$

(**c**) $\Delta\varphi = 60°$ (**d**) $\Delta\varphi = 90°$

(**e**) $\Delta\varphi$ = arbitrary°

Figure 18. Experimental results of the influence of phase difference $\Delta\varphi$ on surface morphology ($\Delta\varphi$).

5. Conclusions

This study proposes a ball-end milling surface morphology prediction model that considers factors, such as tool vibration, eccentricity, and deformation. Based on the theoretical model, the effects of different machining parameters (such as feed rate, line spacing, cutting depth, and phase angle) on surface morphology and roughness were investigated. A validation milling experiment was conducted on quenched steel Cr12MoV workpieces. The main results and conclusions are summarized as follows:

A simulation model based on the Z-map algorithm and MATLAB was developed, and the surface morphology of quenched steel was simulated and analyzed, revealing the formation mechanism of the surface morphology. The simulation results showed good consistency with experimental values. The deviation in three-dimensional surface roughness remained within the range of 7% to 15%, confirming the model's accuracy and reliability. The main reason for the residual deviation is that the simulation assumes ideal conditions without considering factors such as tool axial movement or vibration. In addition, in the actual cutting process, there is a small gap between the tool and the workpiece, which may cause surface scratch and material smearing and increase roughness. Furthermore, tool wear and the rigidity of the machine tool spindle also affect the surface roughness. The combined effects of these factors will lead to surface morphology errors between the simulation and experimental values.

Different machining parameters have a significant impact on the surface morphology of the workpiece. The line spacing (a_e) and feed per tooth (f_z) significantly affect the surface residual height. As the line spacing and feed per tooth increase, the surface residual height and surface roughness also increase markedly. The cutting depth (a_p) has a relatively minor effect on surface morphology, but as the cutting depth increases, the surface roughness also tends to rise. Through simulation analysis, it is observed that the surface morphology of machining presents periodic distribution in the feed path and the interval direction of adjacent tool trajectory.

The phase difference between cutting edges plays a crucial role in influencing the surface morphology of machined surfaces. When the phase difference is at a specific value, the surface texture direction remains consistent, resulting in regular and periodic surface patterns; introducing the phase angle makes the model's predictions closer to actual measurements, providing a theoretical basis and design reference for tool structure optimization and high-quality surface control.

Author Contributions: Conceptualization, Y.C., X.L., M.Z., H.M. and C.L.; Data curation, Y.C., X.L., H.M., Q.L., D.W. and B.Y.; Formal analysis, D.W., B.Y., Q.L. and H.J.; Funding acquisition, Y.C. and X.L.; Investigation, Y.C., M.Z., H.M., Q.L., F.W. and Q.H.; Methodology, C.L.; Project administration, Y.C.; Resources, M.Z. and H.J.; Software, Y.C., X.L., M.Z., C.L., F.W. and Q.H.; Supervision, H.J.; Validation, F.W. and Q.H.; Visualization, D.W. and B.Y.; Writing—original draft, Y.C. and X.L.; Writing—review & editing, Y.C. and X.L. All authors have read and agreed to the published version of the manuscript.

Funding: This research was supported by the General Project of Heilongjiang Postdoctoral Fund (LBH-Z23301); This research was funded by the Research Projects of Basic Scientific Research Business Expenses of Provincial Colleges and Universities in Heilongjiang Province (145409603).

Institutional Review Board Statement: Not applicable.

Informed Consent Statement: Not applicable.

Data Availability Statement: The original contributions presented in this study are included in the article. Further inquiries can be directed to the corresponding author.

Conflicts of Interest: Author Youzheng Cui and Hui Jiang was employed by the company QiQiHar Heavy CNC Equipment Co., Ltd. The remaining authors declare that the research was conducted in the absence of any commercial or financial relationships that could be construed as a potential conflict of interest.

Nomenclature

The key parameters and symbols used in this study are listed below:

a_e	line spacing (radial distance between adjacent cutter paths)
a_p	cutting depth (axial depth of material removal)
f_z	feed per tooth (distance advanced per cutter tooth)
phase difference ($\Delta\varphi$)	phase difference between adjacent cutter teeth (in degrees)
residual height	uncut material height between adjacent scallop tracks

References

1. Du, K.; Li, J.; Pei, Y. Research on hard turning machinability of quenched ultra-high strength steel 45CrNiMoVA. *J. Ordnance Eng.* **2023**, *44*, 773–782.
2. Qin, J.; Xu, X.; Liu, P. Experimental study on surface roughness and residual stress of quenched H13 steel by high-speed milling. *Light Ind. Mach.* **2019**, *37*, 46–51.
3. Guillemot, N.; Lartigue, C.; Billardon, R.; Mawussi, B.K. Prediction of the endurance limit taking account of the microgeometry after finishing milling. *Int. J. Interact. Des. Manuf.* **2010**, *4*, 239–249. [CrossRef]
4. Quinsat, Y.; Lavernhe, S.; Lartigue, C. Surface topography in ball end milling process: Description of a 3D surface roughness parameter. *Mater. Process Technol.* **2008**, *195*, 135–143. [CrossRef]
5. Quinsat, Y.; Lavernhe, S.; Lartigue, C. Characterization of 3D surface topography in 5-axis milling. *Wear* **2011**, *27*, 590–595. [CrossRef]
6. Zhuo, Y.; Han, Z.; An, D.; Jin, H. Surface topography prediction in peripheral milling of thin-walled parts considering cutting vibration and material removal effect. *Int. J. Mech. Sci.* **2021**, *211*, 106797. [CrossRef]
7. Urbikain, G.; Trejo, D.O.; Luo, M.; Lacalle, L.; Elías-Zuiga, A. Surface roughness prediction with new barrel-shape mills considering runout: Modelling and validation. *Measurement* **2021**, *173*, 108670. [CrossRef]
8. Buj-Corral, I.; Vivancos-Calvet, J.; Dominguez-Fernandez, A. Surface topography in ball-end milling processes as a function of feed per tooth and radial depth of cut. *Int. J. Mach. Tools Manuf.* **2012**, *53*, 151–159. [CrossRef]
9. Zhang, X.; Tian, Y.; Li, Z. The influence of milling process parameters on the surface roughness characteristics of tin bismuth alloy and experimental analysis. *Equip. Environ. Eng.* **2019**, *16*, 43–47.
10. Wang, L.; Ge, S.; Si, H.; Yuan, X.; Duan, F. Roughness control method for five-axis flank milling based on the analysis of surface topography. *Int. Mech. Sci.* **2020**, *169*, 105337. [CrossRef]
11. Zhang, W.; Hua, B.; Zhang, L.; Si, B. Modeling and simulation of surface topography in secondary milling with ellipsoid end milling cutter. *Int. J. Interact. Des. Manuf.* **2024**, *18*, 5043–5056. [CrossRef]
12. Wei, J.; Hou, X.; Sun, C. Modeling and simulation of surface topography in five-axis ball end milling. *J. Phys. Conf. Ser.* **2021**, *1820*, 012049. [CrossRef]
13. Sylvain, S.; Quinsat, Y.; Lartigue, C. Model for the prediction of 3D surface topography in 5-axis milling. *Int. J. Adv. Manuf. Technol.* **2010**, *51*, 915–924.
14. Gao, H.; Yue, C.; Liu, X.; Nan, Y. Simulation of surface topography considering cut-in impact and tool flank wear. *Appl. Sci.* **2019**, *9*, 732. [CrossRef]
15. Lei, Z.; Zheng, M.; Zhang, W.; Li, K. Influence and prediction of tool wear on workpiece surface roughness based on milling topography analysis. *Adv. Manuf. Technol.* **2022**, *122*, 1883–1896.
16. Liu, H.; Dong, Y.; Li, Y.; Li, P.; Yang, Z.; Landers, R.G. Geometrical simulation and analysis of ball-end milling surface topography. *Int. J. Adv. Manuf. Technol.* **2019**, *102*, 1885–1900.
17. Layegh, S.E.; Lazoglu, I. 3D surface topography analysis in 5-axis ball-end milling. *CIRP Annals* **2017**, *66*, 133–136. [CrossRef]
18. Felhő, C.; Karpuschewski, B.; Kundrák, J. Surface roughness modelling in face milling. *Procedia Cirp* **2015**, *31*, 136–141. [CrossRef]
19. Yu, Y.; Lin, C.; Hu, Y. Study on simulation and experiment of non-circular gear surface topography in ball end milling. *Adv. Manuf. Technol.* **2021**, *114*, 1913–1923. [CrossRef]
20. Jin, X.; Yu, F.; Wang, X.; Li, Y.; Zhu, Q. Surface topography model of ultra-high strength steel AF1410 based on dynamic characteristics of milling system. *Processes* **2023**, *11*, 641. [CrossRef]

21. Li, Y.; Chen, X.; Wang, Y.; Zhang, W. Surface forming mechanism and process in laser milling of 7075 aluminum alloy with inclination. In Proceedings of the 24th National Laser Conference & Fifteenth National Conference on Laser Technology and Optoelectronics, Shanghai, China, 17–20 October 2020.
22. Lu, X.; Sun, X.; Hou, P.; Xue, L.; Liang, S.Y. Surface topography simulation and roughness prediction of micro-milling single crystal copper. *Int. J. Nanomanuf.* **2021**, *17*, 139–153. [CrossRef]
23. Meng, S. Simulation and Experimental Study on High-Speed Milling Morphology of Cr12MoV Die Steel. Master's Thesis, Harbin University of Science and Technology, Harbin, China, 2020.
24. Sun, M.; Wang, L. Research on the dynamic model of ball end milling cutter and its milling stability. *Mach. Tools Hydraul.* **2012**, *40*, 52–54.
25. Yang, D.; Liu, Z. Surface plastic deformation and surface topography prediction in peripheral milling with variable pitch end mill. *Int. J. Mach. Tools Manuf.* **2015**, *91*, 43–53. [CrossRef]
26. Mei, W. Research on Dynamic Modeling and Simulation System Design of High-Speed Milling Process. Master's Thesis, North China University, Beijing, China, 2011.
27. Cui, Y.; Zheng, M.; Zhang, W.; Wang, B.; Sun, Y.; Wang, W. Study on Fatigue Characteristics of Bionic Functional Surface of Hardened Steel. *Materials* **2020**, *13*, 4130. [CrossRef]

materials

MDPI

Article

Investigation into the Strength, Hydration, and Microstructural Characteristics of Clinker-Free Cement Composed of Phosphorus Slag, Fluidized Bed Combustion Bottom Ash, and Lime

Yanzhou Peng [1,2,*], Haitian Li [2], Hefei Yin [2], Ji Xiao [2,3] and Gang Xu [1,2]

1 Hubei Key Laboratory of Disaster Prevention and Mitigation, China Three Gorges University, Yichang 443002, China; postxg@163.com

2 College of Civil Engineering and Architecture, China Three Gorges University, Yichang 443002, China; 13681221527@163.com (H.L.); yinalbert628@gmail.com (H.Y.); qq375607219@126.com (J.X.)

3 Second Harbor Engineering Company Ltd., Wuhan 430040, China

* Correspondence: pnyzh@163.com

Abstract

This study focuses on developing a novel clinker-free cement, specifically comprising phosphorus slag-based cementitious materials (PSCMs), by utilizing lime and industrial byproducts, including granulated electric furnace phosphorus slag and fluidized bed combustion bottom ash. The optimal composition of PSCM was determined by investigating the effects of different proportions of activators (water glass and sodium sulfate) and retarder (potassium fluoride) on the setting time and the mechanical strength of PSCMs. Performance evaluations demonstrated that the compressive and flexural strengths of the optimal PSCM formulation at 28 days were 64.1 MPa and 7.5 MPa, respectively. Notably, concrete prepared with the optimal PSCM exhibited superior freeze–thaw resistance and sulfate resistance compared to Portland cement concrete of equivalent strength grades. The comprehensive characterization of selected PSCM compositions, conducted using X-ray diffraction (XRD), Fourier transform infrared spectroscopy (FTIR), thermogravimetric analysis (TGA), and scanning electron microscope–energy-dispersive spectrometry (SEM-EDS), provided in-depth insights into the interrelationship among mechanical properties, durability, and microstructural characteristics. SEM-EDS analysis confirmed that calcium aluminosilicate hydrate and sodium aluminosilicate hydrate are the predominant hydration products of PSCMs. FTIR and TG analyses elucidated the continuous hydration behavior of PSCMs during the curing process, while SEM observations revealed a densely compact microstructure in the hardened PSCM paste.

Keywords: phosphorus slag; fluidized bed combustion bottom ash; alkali-activated materials; strength; durability; microstructure

1. Introduction

The growing global population and ongoing urbanization are expected to boost the construction industry, implying there will be consistent demand for building materials such as cement in the foreseeable future. Over the past decade, global cement production has consistently exceeded 4 billion tonnes per year, leading to significant environmental challenges such as carbon dioxide emissions and high energy consumption. Portland cement production, which requires the calcining of limestone at high temperatures, accounts for about 7–8% of global CO_2 emissions [1–3]. Approximately 50–60% of these emissions come from the calcination reaction (a key step in the production process), around 35–40% come from fossil fuel combustion, and come 10–15% from activities such as transportation, space heating, and plant utilities [3–5].

To mitigate CO_2 emissions in cement production, a viable approach is to use industrial solid waste as a partial substitute for cement clinker. This method enables the production of low-clinker cement with comparable or enhanced strength and durability [6,7]. Additionally, using industrial solid waste in clinker-free cement significantly reduces CO_2 emissions and energy consumption. Supplementary cementitious materials (SCMs) such as fly ash (FA), ground granulated blast furnace slag (GGBFS), steel slag, and metakaolin have demonstrated effectiveness in producing clinker-free cement, and are increasingly accepted in the construction market [8–13]. However, the availability of these SCMs, particularly FA and GGBFS, depends on their upstream industries, which may decline in quantity and quality as these industries evolve. Therefore, to support the low-carbon and sustainable development of the cement industry, it is imperative to explore new sources of SCMs for low-clinker or clinker-free cement production.

Granulated electric furnace phosphorus slag (PS), a byproduct of yellow phosphorus production, is generated at a ratio of 8–10 tonnes of PS per tonne of yellow phosphorus [14,15]. PS primarily consists of SiO_2 and CaO, with an amorphous glass phase content reaching up to 85% by weight, which confers significant pozzolanic reactivity suitable for cementitious materials [16–19]. However, the low Al_2O_3 content and high SiO_2 concentration limit its early reactivity [20–26]. Moreover, the presence of P_2O_5 and F in PS retards the hydration of Portland cement, adversely affecting early strength development [27,28]. Consequently, the utilization of PS remains limited, resulting in substantial stockpiles and the environmental contamination of soil and water. Therefore, further research into effective utilization strategies for PS is urgently required.

Significant efforts have been undertaken to enhance the reactivity of phosphorus slag (PS) through increasing its specific surface area, elevating curing temperatures (including steam curing), and incorporating chemical activators [29–32]. These approaches have not only improved PS utilization, but also promoted resource conservation and environmental protection. However, challenges persist, particularly concerning the high energy consumption during grinding and the elevated costs associated with high-temperature curing.

Research findings [33–36] indicate that aluminosilicate materials such as fly ash, GGBFS, and metakaolin, when combined with alkaline components, can produce alkali-activated materials (AAMs), characterized by low carbon emissions and reduced energy consumption. These AAMs exhibit superior chemical and freeze–thaw resistance compared to Portland cement [37–40], positioning them as promising energy-efficient and environmentally friendly building materials [8–13,41]. Phosphorus slag can also be utilized to produce AAMs [42–46]. Extensive studies have explored the development and application of these PS-based AAMs, yielding significant results [47,48]. However, research on alkali-activated PS materials highlights several challenges [18,20,22,24,41,47,49]. First, their production often requires blending PS with Portland cement, GGBFS, fly ash, or metakaolin, leading to a relatively low proportion of PS. Second, concerning fluidized bed combustion bottom ash (BA), the presence of which in emissions has exhibited a steady increase over recent years [50], there are currently no public reports regarding its application in the preparation of alkali-activated materials (AAMs).

This study proposes a novel clinker-free cement composed of phosphorus slag-based cementitious material (PSCM), developed by utilizing lime and industrial solid wastes, including granulated electric furnace phosphorus slag (PS) and fluidized bed combustion bottom ash (BA). This approach promotes the recycling and reuse of these materials. The effects of varying dosages of activators (water glass and sodium sulfate) and retarder (potassium fluoride) on the PSCM's properties (setting time and strength) were first investigated to determine the optimal composition. Additionally, the sulfate and freeze–thaw resistances of PSCM concrete were examined. Finally, the hydration and microstructure of selected PSCM compositions were analyzed using X-ray diffraction (XRD), Fourier transform infrared spectroscopy (FTIR), thermogravimetry (TG), and scanning electron microscopy (SEM). This paper aims to achieve the following objectives:

(1) To develop PSCMs with a strength grade of 42.5 using phosphorus slag and fluidized bed combustion bottom ash;
(2) To evaluate the freeze–thaw resistance and sulfate resistance of concrete prepared with the developed PSCMs;
(3) To investigate the relationship between the performance and microstructure of the hardened paste of the developed PSCMs through microstructural analysis.

2. Experimental Programs

2.1. Materials

Ground granulated electric furnace phosphorus slag (PS), fluidized bed combustion bottom ash (BA), and lime (L) were utilized as the primary ingredients of phosphorus slag-based cementitious materials (PSCMs) in this study. The materials used in this study are as follows: Granulated electric furnace phosphorus slag (PS), produced by Hubei Yatai Chemical Co., Ltd. (Yichang, China), has a specific surface area of 397 m^2/kg. Circulating fluidized bed combustion bottom ash (BA), supplied by Hubei Yihua Group Co., Ltd. (Yichang, China), is of the low-sulfur type and has a specific surface aera of 579 m^2/kg. Lime (L) was provided by Hubei Yichang Runchu Building Materials Co., Ltd. (Yichang, China). The slaking time and slaking temperature of the lime are 7 min and 60 °C, respectively. The chemical compositions are detailed in Table 1, and the XRD patterns of PS powder and BA are shown in Figure 1. As illustrated in Figure 1, the amorphous glass phase is the primary mineralogical component of PS, indicating its high reactivity [18,27,37]. In contrast, BA is primarily composed of quartz with a minor amount of mullite. ISO standard silica sand, compliant with Chinese Standard GB/T 17671-2021 [51], was used to prepare test specimens for evaluating PSCM performance. River sand with a fineness modulus of 2.5 and crushed limestone (5–20 mm) served as aggregates in the PSCM concrete.

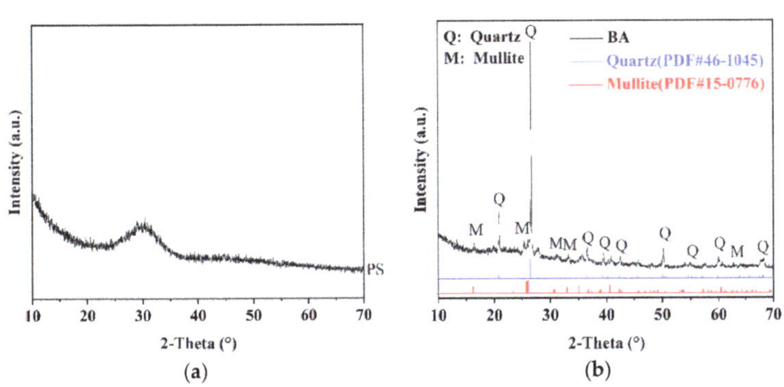

Figure 1. XRD patterns of industrial solid waste used in this study: (**a**) ground granulated electric furnace phosphorus slag (PS), and (**b**) fluidized bed combustion bottom ash (BA).

Table 1. Chemical composition of phosphorus slag (PS), fluidized bed combustion bottom ash (BA), and L (wt. %).

	SiO_2	Al_2O_3	Fe_2O_3	CaO	MgO	K_2O	Na_2O	SO_3	P_2O_5	Loss on Ignition
PS	38.09	3.85	0.33	43.64	1.98	3.05	0.37	1.27	4.21	0.52
BA	55.74	27.52	3.85	2.26	1.59	1.36	0.51	1.35	0.22	3.69
L	0.44	0.18	0.07	93.4	0.95	/	/	0.79	/	4.10

Meanwhile, water glass and sodium sulfate were employed as activators for PSCMs. The water glass used was a commercially sourced aqueous solution with a modulus of 1.2, whereas the sodium sulfate was a commercially available analytical reagent. Furthermore, potassium fluoride (KF) of analytical grade was added as a retarder to modulate the setting time of alkali-activated PSCMs.

2.2. Mix Proportions and Samples Preparation

The formulations of the PSCM mixtures considered in this study are tabulated in Table 2.

Table 2. Formulations of PSCM mixtures.

Mixture No.	Cementitious Component (%)			Activator [a]		Retarder [a] (i.e., KF) (%)
	PS	BA	Lime (L)	Water Glass (WG) (Na_2O wt. %)	Sodium Sulfate (SS) (wt. %)	
BA0-WG5	95	0	5.0	5.0	0	1.0
BA0-WG6	95	0	5.0	6.0	0	1.0
BA0-WG7	95	0	5.0	7.0	0	1.0
BA0-WG8	95	0	5.0	8.0	0	1.0
BA15-WG5	80	15	5.0	5.0	0	1.0
BA15-WG6	80	15	5.0	6.0	0	1.0
BA15-WG7	80	15	5.0	7.0	0	1.0
BA15-WG8	80	15	5.0	8.0	0	1.0
BA0-WG7-SS1	95	0	5.0	7.0	1.0	1.0
BA0-WG7-SS3	95	0	5.0	7.0	3.0	1.0
BA0-WG7-SS5	95	0	5.0	7.0	5.0	1.0
BA15-WG7-SS1	80	15	5.0	7.0	1.0	1.0
BA15-WG7-SS3	80	15	5.0	7.0	3.0	1.0
BA15-WG7-SS5	80	15	5.0	7.0	5.0	1.0
BA15-WG7-SS3-KF0	80	15	5.0	7.0	3.0	0.0
BA15-WG7-SS3-KF0.5	80	15	5.0	7.0	3.0	0.5
BA15-WG7-SS3-KF1	80	15	5.0	7.0	3.0	1.0
BA15-WG7-SS3-KF1.5	80	15	5.0	7.0	3.0	1.5

[a] The dosages of admixtures, including activator and retarder, are expressed as percentages of the total mass of PS, BA, and L.

PSCM paste, mortar, and concrete were prepared according to the formulations presented in Table 2. A 5 L capacity mortar mixer was utilized to mix the PSCM mortar with a water-to-binder ratio (W/B) of 0.5, including water from the activator solution, and a sand-to-binder ratio of 3. Cementitious components (PS, BA, and L) and ISO standard silica sand were initially mixed for 30 s. The alkaline activator solution and mixing water were then gradually added until a uniform mortar was achieved. Fresh mortar was immediately cast into prisms measuring 40 mm × 40 mm × 160 mm and covered with plastic film to prevent moisture evaporation. After 24 h, the specimens were demolded and cured at a temperature of 20 ± 2 °C and a relative humidity of 95% until they reached the designated testing age.

The preparation procedure for the PSCM paste assessed for setting time closely followed that of the PSCM mortar. For the PSCM concrete samples the mix proportions of which are shown in Table 3, all components—including cementitious components (PS, BA, and L), aggregates (river sand and crushed limestone), retarder, alkaline activator solution, and mixing water—were mixed using a 60 L mixer. The thoroughly mixed PSCM was cast and vibrated as per conventional concrete practices. Cubic specimens (150 mm × 150 mm × 150 mm and 100 mm × 100 mm × 100 mm) as well as the prismatic specimen (100 mm × 100 mm × 400 mm) were prepared to evaluate compressive strength, sulfate attack resistance, and freeze–thaw resistance, respectively. After demolding at for 24 h, these specimens were cured at 20 ± 2 °C and 95% until testing.

Table 3. The mix proportions of PSCM concrete.

Concrete No.	W/B	Mix Design (kg/m^3)			Activator [c]		Retarder [c]
		PSCM	Sand	Crushed Stone	Water Glass (WG) (Na_2O wt. %)	Sodium Sulfate (SS) (wt. %)	KF (wt. %)
PSCM-C30	0.44	333 [a]	746	1121	7	3	1
PSCM-C50	0.33	500 [a]	578	1122	7	3	1
OPC-C30	0.44	333 [b]	746	1121	0	0	0
OPC-C50	0.33	500 [b]	578	1122	0	0	0

[a] The mass ratio of the components in PSCM, namely, PS, BA, and L, is 80:15:5. [b] Ordinary Portland cement concrete of strength grades C30 and C50 (OPC-C30 and OPC-C50), prepared using P·O 42.5 Portland cement, served as the reference groups in this study. This cement, supplied by China Huaxin Cement Co., Ltd. (Wuhan, China), exhibited compressive strengths of 31.4 MPa at 3 days and 47.5 MPa at 28 days. [c] The dosages of admixtures, including activator and retarder, are expressed as a percentage of the mass of the PSCM.

2.3. Methods

Figure 2 illustrates the testing methodology employed in this study.

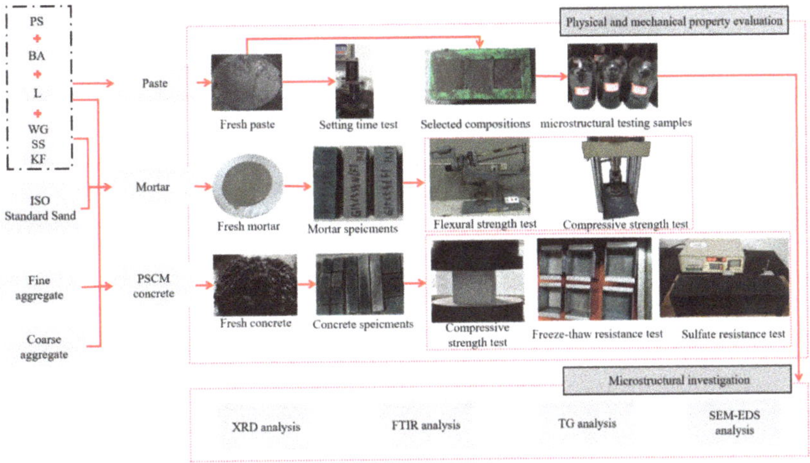

Figure 2. Testing methodology of study.

2.3.1. Determination of the Setting Time and Strength of PSCMs

The setting time of PSCMs was determined using a standard consistency paste in accordance with GB/T 1346-2024 [52], utilizing a Vicat apparatus equipped with a needle designed to penetrate the samples with paste at regular intervals.

The strength of PSCMs was evaluated using prismatic mortar specimens measuring 40 mm × 40 mm × 160 mm. These prismatic specimens were prepared as described in Section 2.2. The flexural and compressive strength were measured at 3 and 28 days, following the testing protocol prescribed by GB/T 17671-2021 [51]. Specifically, the flexural strength was calculated as the average of the measurements obtained from three prism specimens (40 mm × 40 mm × 160 mm), while the compressive strength was determined as the arithmetic mean of the results for six specimens.

2.3.2. Determination of PSCM Concrete Performance

The compressive strength of PSCM concrete, with mix proportions as detailed in Table 3, was evaluated at 3 and 28 days according to GB/T 50081-2019 [53]. Additionally, tests for sulfate attack resistance and freeze–thaw resistance were conducted following the methods specified in GB 50082-2024 [54], namely, the Test Method for Rapid Freezing and Thawing and the Test Method for Resistance of Concrete to Sulfate Attack.

2.3.3. Microstructure Examination

The mixtures BA0-WG7-KF1, BA0-WG7-SS3-KF1, BA15-WG7-KF1, and BA15-WG7-SS3-KF1 of the PSCM paste, as presented in Table 2, were selected for microstructural examination. Cubic specimens with dimensions of 40 mm × 40 mm × 40 mm were prepared according to the formulations shown in the table, using a water-to-binder ratio (W/B) of 0.30.

For each mixture, fresh paste was cast into cubic molds and immediately covered with plastic film to prevent moisture loss. After 24 h, the specimens were demolded and subsequently cured at 20 ± 2 °C and 95% until the ages of 3 days and 28 days. Samples were taken for microstructural investigation from the central sections of the cubic specimens at the designated testing age. These samples were then immersed in ethanol for at least 24 h to facilitate dehydration and subsequently dried in an oven at 40 °C for an additional 24 h. Samples intended for XRD, FTIR, and TG-DTG analysis were ground to fine powders with particle sizes less than 10 μm, while those prepared for SEM observations were trimmed to approximate dimensions of 3 mm × 3 mm × 1 mm.

XRD analysis was conducted using an AXS D8-Advance X-ray diffractometer (Bruker, Karlsruhe, Germany) with a scan range of 5° to 70° (2θ) and a step size of 0.02°. FTIR analysis was performed utilizing a Thermo Nicolet Fourier transform infrared spectroscopy (PerkinElmer, Waltham, MA, USA), covering a wavenumber range from 4000 cm^{-1} to 400 cm^{-1}. Thermogravimetric analysis (TGA) was conducted using a Netzsch STA 449F3 thermogravimetric analyzer (Netzsch, Bavaria, Germany) under an air atmosphere with the temperature increasing from 20 °C to 1000 °C at a heating rate of 3 °C/min. SEM-EDS analysis was performed using a JSM-IT800 environmental scanning electron microscope (JEOL, Tokyo, Japan) at an acceleration voltage of 20 kV.

3. Experimental Results and Discussion

3.1. Effect of Activator on the Strength of PSCM

Based on previous research [28], two formulations of clinker-free phosphorus slag-based cementitious materials (PSCMs) were selected: BA0 (0% BA, 5% lime) and BA15 (15% BA, 5% lime). This section experimentally investigates the effects of water glass (WG) and sodium sulfate (SS) dosages on the compressive and flexural strength of PSCMs. WG dosages, defined as the mass ratio of Na2O from WG to the total cementitious materials (PS + BA + L), were set at 5%, 6%, 7%, and 8%. SS dosages, expressed as the mass ratio of SS to the total cementitious materials, were set at 1%, 3%, and 5%. Detailed formulations are provided in Table 2.

3.1.1. Effect of the Water Glass Dosage

The flexural and compressive strengths of mixtures BA0 and BA15, with water glass dosages of 5%, 6%, 7%, and 8%, are diagrammed in Figures 3 and 4, respectively.

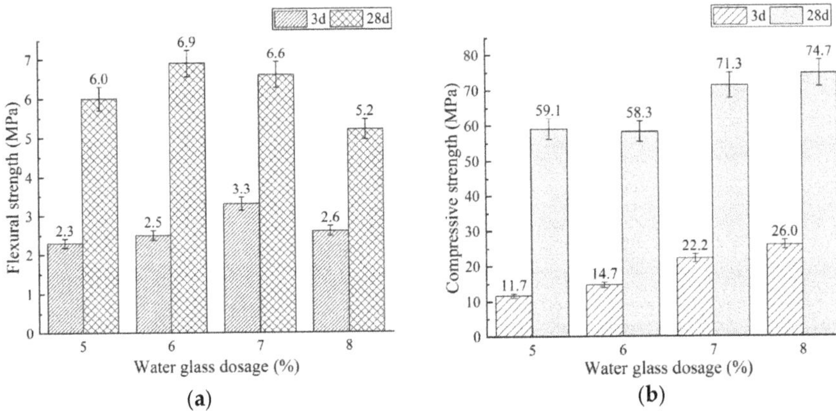

Figure 3. Effect of water glass dosage on strength of mixture BA0: (**a**) flexural strength and (**b**) compressive strength.

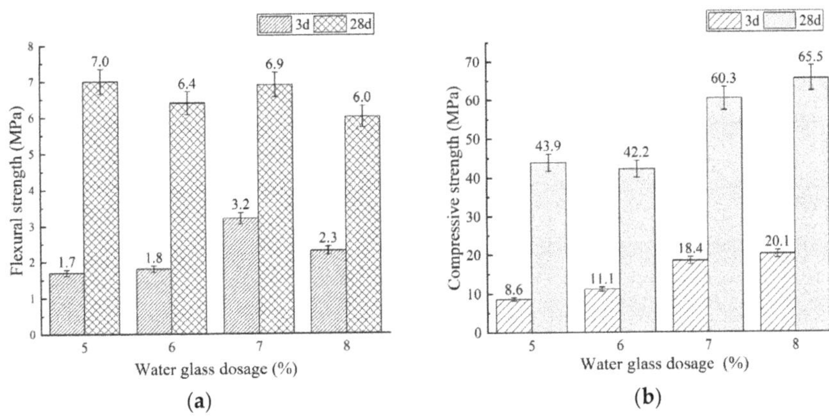

Figure 4. Effect of water glass dosage on strength of mixture BA15: (**a**) flexural strength and (**b**) compressive strength.

As shown in Figures 3 and 4, increasing the water glass dosage from 5% to 8% resulted in a consistent increase in compressive strength for both BA0 and BA15 PSCM mixtures (Figures 3b and 4b). Flexural strength initially increased but then decreased (Figures 3a and 4a). Notably, at a 7% water glass dosage, the 3-day flexural strengths of BA0 and BA15 reached peak values of 3.3 MPa and 3.2 MPa, respectively. In fact, the formation of alkali-activated materials involves the geopolymerization of aluminosilicate constituents in raw materials (precursors) at room temperature or high temperatures [55–59]. This process can be divided into three stages: the depolymerization stage, the repolymerization stage, and the hardening stage [56–59]. Specifically, the steps are as follows: (1) Aluminosilicate components in precursor are dissolved in the alkaline activator solution, releasing free silicon–oxygen and aluminum–oxygen tetrahedral units. (2) These tetrahedral units repolymerize through hydroxyl condensation reactions, and (3) the silicoaluminate network structure gradually hardens, forming calcium aluminosilicate hydrate (C-A-S-H) and sodium aluminosilicate hydrate (N-A-S-H) gels. Therefore, higher water glass dosages increased the concentrations of OH^-, silicon–oxygen tetrahedral units, and aluminum–oxygen tetrahedral units, thereby promoting the geopolymerization reactions of phosphorus slag-based materials. This enhancement significantly improved mechanical properties, particularly compressive strength.

Furthermore, Figure 4 shows that under identical water glass dosage conditions, mixture BA15 exhibits lower 3-day flexural and compressive strengths, as well as 28-day compressive strength, compared to mixture BA0. This indicates that incorporating fluidized bed combustion bottom ash (BA) reduces the early-age strength of alkali-activated phosphorus slag-based materials. This phenomenon is closely associated with the chemical compositions of BA and PS: BA is a low-calcium silicon–aluminum material, while PS is high-calcium (Table 1). Generally, alkali-activated high-calcium silicon–aluminum materials, such as ground granulated blast furnace slag (GGBFS), primarily form C-A-S-H hydration products with high calcium contents. In contrast, alkali-activated low-calcium silicon–aluminum materials, including metakaolin and fly ash, predominantly generate N-A-S-H-based products [55,56,60]. At room temperature, the reaction rate of alkali activation for the low-calcium silicon–aluminum material is relatively slow, leading to extended setting times and reduced early-age strength. Consequently, the incorporation of BA, which has a lower calcium oxide content, resulted in a significant reduction in the strength of PSCMs, especially at 3 days. However, when the water glass content was 7%, both Mixtures BA0 and BA15 met the strength requirements for 42.5-grade Portland Phosphorus slag cement specified in JC/T 740-2006 [25], with the exception of the 3-day flexural strength, which was slightly below the threshold (not less than 3.5 MPa).

Additionally, as can be observed from Figures 4 and 5, the flexural and compressive strengths of PSCMs exhibit distinct variation trends with increasing WG dosages. Specifically, the compressive strength increases steadily as the WG content rises, whereas the flexural strength initially increases and subsequently decreases. This divergence can be attributed to the different sensitivities of compressive strength and flexural strength to microstructural characteristics; the compressive strength is primarily influenced by overall compactness, while the flexural strength is more sensitive to the uniformity of pore distribution [61]. At low WG contents, the enhanced geopolymerization reaction generates additional hydration products, thereby increasing the overall compactness of the hardened paste and promoting both compressive and flexural strengths. However, when the WG content exceeds a critical threshold, the microstructural uniformity deteriorates, leading to uneven pore distribution and a reduction in flexural strength.

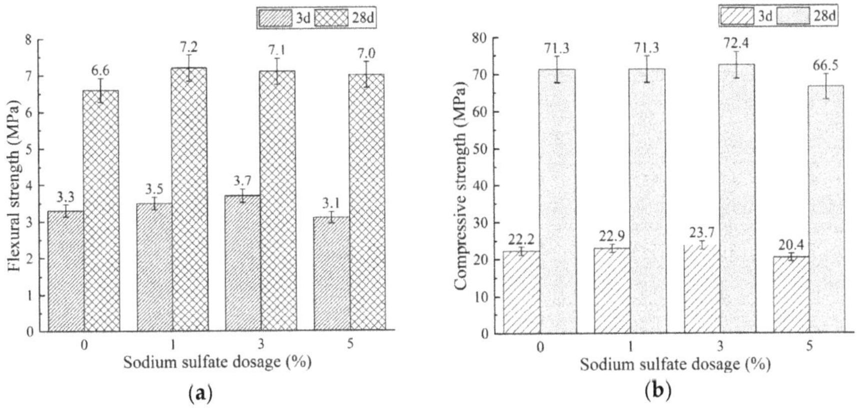

Figure 5. Effect of sodium sulfate (SS) dosage on strength of mixture BA0-WG7: (**a**) flexural strength and (**b**) compressive strength.

3.1.2. Effect of the Sodium Sulfate Dosage

Sodium sulfate (SS) was introduced as an additional activator to enhance the early-age flexural strengths of Mixtures BA0-WG7 and BA15-WG7, both of which contain 7% water glass (as detailed in Table 2). The effects of varying dosages (1%, 3%, and 5%) of SS on the compressive and flexural strengths of these mixtures were examined. The results are shown in Figures 5 and 6.

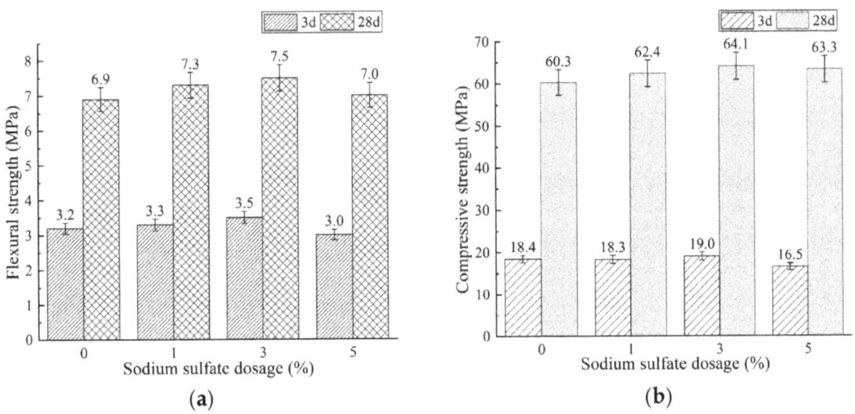

Figure 6. Effect of sodium sulfate (SS) dosage on the strength of mixture BA15-WG7: (**a**) flexural strength and (**b**) compressive strength.

As shown in Figures 5a and 6a, the addition of sodium sulfate (SS) enhanced the flexural strengths of mixtures BA0-WG7 and BA15-WG7. Specifically, for mixture BA0-WG7, the 3-day flexural strength increased from 3.3 MPa (without SS) to 3.5 MPa and 3.7 MPa with the addition of 1% and 3% SS, respectively, representing increases of 6.1% and 12.1%. The 28-day flexural strength also rose from 6.6 MPa (without SS) to 7.2 MPa and 7.1 MPa. For mixture BA15-WG7, the 3-day flexural strength increased from 3.2 MPa (without SS) to 3.3 MPa and 3.5 MPa with 1% and 3% SS, while the 28-day flexural strength increased from 6.9 MPa (without SS) to 7.3 MPa and 7.5 MPa. However, a further increase in SS content to 5% resulted in a decrease in flexural strength for both mixtures. Additionally,

Figures 5b and 6b show that the effect of SS on the compressive strengths of mixtures BA0-WG7 and BA15-WG7 follows a similar trend to its effect on flexural strength, but with smaller changes in magnitude. This was because the composite activation of WG and SS accelerated the hydration rate of phosphorus slag-based materials (namely, mixture BA0-WG7 and BA15-WG7), thereby shortening their setting time. Additionally, besides C-A-S-H and N-A-S-H gels, a certain amount of monosulfate was also produced [61–63], which enhanced the early-age strength, particularly the flexural strength. However, when the SS dosage exceeded 3%, the paste exhibited rapid setting, affecting the morphology and uniformity of the hydration products, and adversely impacting the uniformity and porosity of the hardened paste, ultimately leading to a reduction in strength.

Figures 5 and 6 demonstrate that mixtures BA0-WG7-SS3 and BA15-WG7-SS3 both comply with the compressive and flexural strength requirements stipulated in JC/T 740-2006 [25] for P.PS 42.5 Portland Phosphorous Slag Cement at both 3 and 28 days. To promote the utilization of diverse industrial solid wastes in low-carbon cementitious materials, subsequent experiments focus on the mixture BA15-WG7-SS3. We then investigate the regulation of setting time for this mixture, followed by the preparation of PSCM concrete. Subsequently, experimental studies evaluate the freeze–thaw resistance and sulfate resistance of the PSCM concrete.

3.2. *Effect of Retarder Dosage on the Setting Time and Strength of PSCMs*

During the investigation of the effect of sodium sulfate dosage on mixture BA15-WG7-SS3, it was observed that this PSCM mixture exhibits a relatively rapid setting time. To address this, potassium fluoride (KF) was introduced as a retarder to adjust the setting time. This section studies the effects of varying KF dosages on the initial and final setting times, as well as the compressive and flexural strengths of the mixture. The results are summarized in Figures 7 and 8.

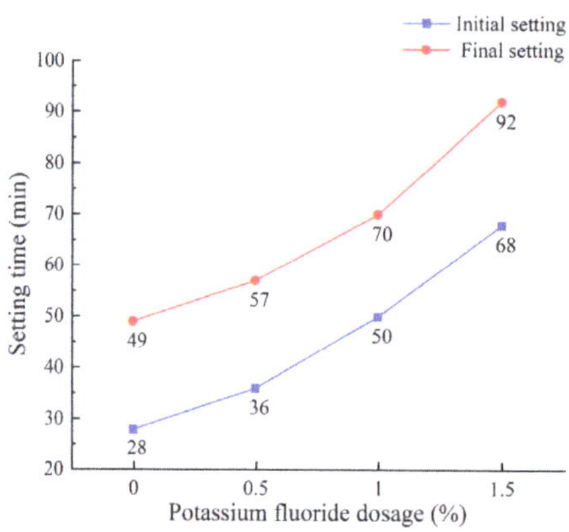

Figure 7. Effect of KF dosage on setting time of mixture BA15-WG7-SS3.

As shown in Figure 7, the setting time of PSCM increased with a higher KF dosage. This is due to the reaction between fluoride ions (from KF) and calcium ions (from PS, BA, and L) during the early stages of hydration, leading to the formation of CaF2, which has extremely low solubility. The deposition of CaF2 on the surface of unhydrated PSCMs particles inhibits their further dissolution and subsequent hydration, thereby prolonging the setting time [64]. As the KF dosage increases, this inhibitory effect becomes more pronounced, allowing KF to effectively regulate the setting time of PSCMs. However, excessive KF dosages (e.g., 1.5%) result in a significant retarding effect, thereby reducing the strength of PSCMs (Figure 8).

Figure 8 indicates that as the KF dosage increases from 0% to 1.0%, both the flexural and compressive strengths of mixture BA15-WG7-SS3 gradually improve. Specifically, at 3 days, the flexural strength rises from 1.8 MPa to 3.5 MPa, and the compressive strength increases from 10.0 MPa to 19.0 MPa; at 28 days, the flexural strength increases from 6.5 MPa to 7.5 MPa, and the compressive strength rises from 38.9 MPa to 64.1 MPa. However, when the KF dosage further increases to 1.5%, the PSCM strength begins to decline. At this point, the 28-day flexural and compressive strengths are 6.3 MPa and 53.7 MPa, respectively. This can be attributed to the influence of the retarder on the depolymerization–repolymerization–hardening process of AAMs [56,58,64]. Specifically, when the mixture BA15-WG7-SS3 is mixed with water, free silicon–oxygen tetrahedral units and aluminum–oxygen tetrahedral units, along with calcium ions, are derived via dissolution from phosphorus slag (PS) and fluidized bed combustion bottom ash (BA) under the combined activating effects of water glass and sodium sulfate. Subsequently, these ions undergo polycondensation reactions to form gel products such as C-A-S-H and N-A-S-H gels, which gradually harden.

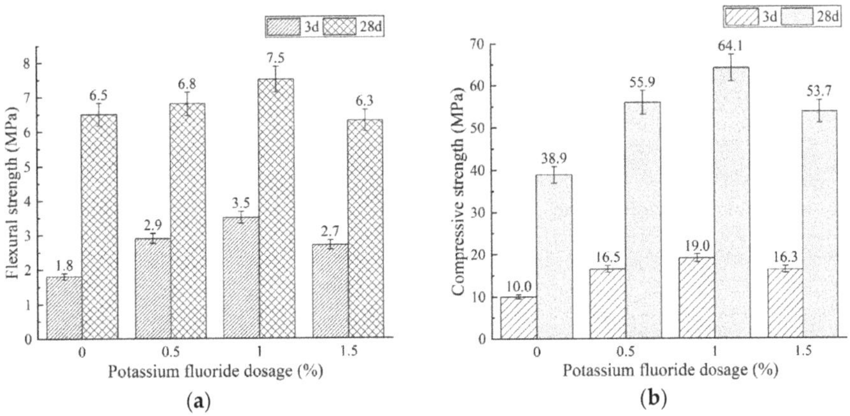

Figure 8. Effect of potassium fluoride dosage on strength of mixture BA15-WG7-SS3: (**a**) flexural strength and (**b**) compressive strength.

In the absence of KF, the reaction rates at each stage are significantly faster, leading to a relatively porous microstructure of the formed products and consequently lower strength in the hardened paste (Figure 8, KF% = 0). However, with the incorporation of KF, insoluble CaF_2 forms on the surfaces of cementitious material particles, inhibiting the dissolution of silicon–oxygen and aluminum–oxygen tetrahedra units in the precursor. This delays the formation and growth of products such as C-A-S-H and N-A-S-H gels, resulting in a more improved microstructure of the hydration products and a reduction in the total porosity of the hardened paste, thereby enhancing its strength. When the KF dosage is 1.0%, the strength of the hardened paste reaches its peak, with 3-day flexural and compressive strengths of 3.5 MPa and 19.0 MPa, respectively. However, when the KF dosage increases to 1.5%, the excessively strong retarding effect reduces the quantity of C-A-S-H and N-A-S-H gels, leading to a decline in the hardened paste strength. Nevertheless, compared to the samples lacking KF, here, the microstructure at 3 days and 28 days was more uniform and denser, resulting in higher strength.

Moreover, Figures 7 and 8 demonstrate that at a KF dosage of 1.0%, such as in mixture BA15-WG7-SS3, meets the setting time and strength requirements specified in JC/T 740-2006 [25] for P.PS 42.5-grade Portland Phosphorous Slag Cement. Therefore, the optimal PSCM formulation was determined to be BA15-WG7-SS3-KF1.0, which was used to prepare a PSCM concrete that would be evaluated for freeze–thaw resistance and sulfate attack resistance.

3.3. *Freeze–Thaw Resistance and Sulfate Attack Resistance of PSCM Concrete*

3.3.1. Freeze–Thaw Resistance

The freeze–thaw resistance results of PSCM concrete are shown in Table 4. Figure 9 presents the appearances of PSCM-C50 and OPC-C50 specimens after 175 freeze–thaw cycles.

Table 4. Freeze–thaw resistance results of PSCM concrete.

Concrete No.	Mass Loss of Specimens After n Freeze–Thaw Cycles, ΔW_n/%					Relative Dynamic Elastic Modulus of Specimens After n Freeze–Thaw Cycles, P/%				
	0	**50**	**100**	**150**	**175**	**0**	**50**	**100**	**150**	**175**
PSCM-C30	0	0	0.5	/	/	100	88.1	64.2	/	/
PSCM-C50	0	0	0	0.08	0.10	100	91.2	83.1	68.3	61.1
OPC-C30	0	0.4	1.7	/	/	100	79.1	53.5	/	/
OPC-C50	0	0	0.1	1.24	2.07	100	96.2	81.7	62.8	56.6

Table 4 shows that after 100 freeze–thaw cycles, although the mass loss rates of both PSCM-C30 and OPC-C30 specimens were below 5%, the relative dynamic elastic modulus of PSCM-C30 was 64.2%, while that of OPC-C30 decreased to 53.5% (below the threshold of 60%). This indicates that PSCM-C30 exhibited superior freeze–thaw resistance compared to OPC-C30, achieving resistance grades of F100 and F50, respectively. For C50 concrete, after 150 freeze–thaw cycles, the mass loss rates for PSCM-C50 and OPC-C50 were 0.08% and 1.24%, respectively, and their relative dynamic elastic moduli decreased to 62.8% and 68.3%, respectively. Both achieved a freeze–thaw resistance grade of F150. However, after extending the freeze–thaw cycles to 175 (Figure 9), although the mass loss rates of both specimens remained below 5%, the relative dynamic elastic modulus of OPC-C50 dropped to 56.6%, while that of PSCM-C50 remained above 61.1%. In conclusion, at the same strength grade, the PSCM concrete demonstrated a slightly better freeze–thaw resistance than the Portland cement concrete.

(a)

(b)

Figure 9. Appearance of PSCM concrete specimens after 175 freeze–thaw cycles: (**a**) PSCM-C50 and (**b**) OPC-C50 specimens.

3.3.2. Sulfate Attack Resistance

The sulfate attack resistance results for PSCM concrete are displayed in Table 5.

Table 5. Sulfate attack resistance results of PSCM concrete.

Concrete No.	Compressive Strength Corrosion Resistance Coefficient of PSCM Concrete After *n* Cycles of Dry–Wet Testing with Sodium Sulfate Solution, K_f (%)		
	30	60	90
PSCM-C30	110	99	85
PSCM-C50	121	112	103
OPC-C30	93	71	/
OPC-C50	104	95	74

As shown here, after 30 cycles of dry–wet testing with a sodium sulfate solution, the compressive strength corrosion resistance coefficients for all concrete specimens, except OPC-C30, exceeded 100%. This indicates that during the initial stage of sulfate attack, these specimens exhibited higher compressive strength values compared to standard-cured specimens of the same age. However, as the numbers of dry–wet cycles increased to 60 and 90, the coefficients significantly decreased. This phenomenon is due to the simultaneous early-stage hydration of PSCMs and sulfate attack [23,47,49]. Initially, newly formed hydration and sulfate attack products such as monosulfate filled the pores of the hardened paste, improving pore structure and reducing porosity, thereby enhancing compressive strength. However, prolonged erosion and intensified attack led to an accumulation of erosion products, resulting in micro-crack formation and propagation within the concrete matrix, which adversely affected the sample's mechanical properties [18,24].

Furthermore, Table 5 demonstrates that the corrosion resistance coefficients for PSCM-C30 concrete were 99% and 85% after 60 and 90 dry–wet cycles, respectively, while that for OPC-C30 was only 71% after 60 cycles (below the threshold of 75%). This indicates that the sulfate resistance grade of PSCM-C30 is KS90, whereas that of OPC-C30 is rated at KS30. Similarly, for PSCM-C50 and OPC-C50, the corrosion resistance coefficients were 103% and 74%, respectively, after 90 dry–wet cycles, indicating that the sulfate resistance grade of PSCM-C50 is higher than that of KS90, while that of OPC-C50 is rated at KS60. These results suggest that, at the same strength grade, PSCM concrete exhibited significantly higher sulfate resistance than Portland cement concrete. This enhanced performance may be attributed to the activating effect of sulfate on PSCMs.

3.4. Microstructural Investigation on PSCM

3.4.1. XRD Analysis

Based on the aforementioned findings, the optimal proportion of PSCMs was selected to investigate hydration hardening behavior and microstructural characteristics. Moreover, three specific PSCM compositions, namely, mixtures BA0-WG7-KF1, BA0-WG7-SS3-KF1, and BA15-WG7-KF1, were additionally chosen. By systematically analyzing the effects of the presence or absence of BA and SS on the hydration processes and microstructural features of PSCM, the intrinsic relationships between material composition, microstructural characteristics, and macroscopic properties (e.g., mechanical strength and durability) of PSCMs were thoroughly explored.

Figure 10 illustrates the XRD patterns of samples derived from the pastes of mixtures BA0-WG7-KF1, BA0-WG7-SS3-KF1, BA15-WG7-KF1, and BA15-WG7-SS3-KF1, as listed in Table 2.

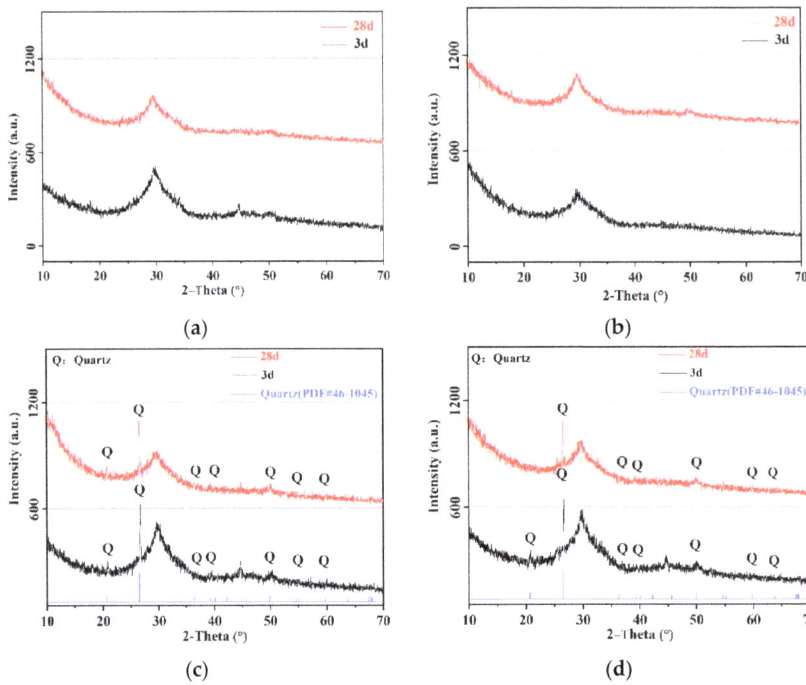

Figure 10. XRD patterns of paste samples from the selected PSCM compositions: (a) BA0-WG7-KF1, (b) BA0-WG7-SS3-KF1, (c) BA15-WG7-KF1, and (d) BA15-WG7-SS3-KF1.

As shown in Figure 10, the four PSCM compositions display a broad diffraction peak within the 25° to 35° range of their XRD patterns. Additionally, diffraction peaks corresponding to quartz crystals were observed in the XRD patterns of the two PSCM compositions containing fluidized bed combustion bottom ash (BA) (Figure 10c,d), while no other crystalline phases were detected. Based on the chemical compositions of PS and BA (Table 1), as well as the XRD analysis results (Figure 1), it can be inferred that the broad diffused peak between 25° and 35° originated from incompletely hydrated PS, while the quartz crystals were introduced by BA. These findings indicate that the main hydration products of alkali-activated PSCMs are amorphous substances, specifically C-A-S-H and N-A-S-H gels, with no significant crystalline phases detected. This is consistent with the subsequent SEM analysis results presented in Section 3.4.4. It can also be seen in Figure 10 that the intensities of diffraction peaks within the range of 25–35° in the XRD patterns for these four PSCM compositions at 28 days are weaker compared to those of the 3 day samples. This observation suggests that PS is progressively consumed as the curing time extends. This consumption facilitates the formation of additional amorphous hydration products, thereby significantly enhancing the strength of the hardened paste. Furthermore, comparing Figure 10a with Figures 10b and 10c with Figure 10d reveals that the addition of sodium sulfate does not significantly alter the types of main products in the PSCMs.

3.4.2. FTIR Analysis

Figure 11 presents the FTIR spectra of the selected PSCM samples, cured for 3 days and 28 days. The consistency of the spectra across all samples indicates that the incorporation of fluidized bed combustion bottom ash (BA) or sodium sulfate has minimal impact on the primary hydration products in water glass-activated PSCMs during both early and late hydration stages.

In the FTIR analysis of cementitious materials, vibrational absorption bands with wavenumbers below 1000 cm^{-1} primarily correspond to silicates and carbonates in the hydration products, while those above 1600 cm^{-1} are typically attributed to chemically bound water [65–74]. Specifically, the bands at approximately 460 cm^{-1}, 670 cm^{-1}, and 975 cm^{-1} are associated with the vibration of Si–O–T bonds (where T represents tetrahedral Si or Al), the bands at 1422 cm^{-1} and 1479 cm^{-1} correspond to the vibration of O–C–O bonds in CO_3^{2-}, and the bands at 1640 cm^{-1} and 3440 cm^{-1} are attributed to chemically bound water. The vibrational absorption bands corresponding to Si–O–T, O–C–O, and H–O–H bonds are identified in the spectra (Figure 11).

As shown in Figure 11, two broad and intense bands at approximately 459 cm^{-1} and 978 cm^{-1}, along with a weak band at 669 cm^{-1}, are observed in the spectra of all samples. These bands are attributed to the bending, asymmetric stretching, and symmetric stretching vibrations of Si–O–T bonds, respectively [70–72]. The presence of these characteristic vibration bands indicates the formation of C-A-S-H and N-A-S-H gels in the hydration products of the PSCM compositions. Moreover, the intensity of the two broad intense bands at approximately 459 cm^{-1} and 978 cm^{-1} increases with curing age, suggesting increases in the contents of C-A-S-H and N-A-S-H gels over time. Meanwhile, the band at 669 cm^{-1} becomes sharper as curing progresses, indicating the enhanced structural ordering of the C-A-S-H and N-A-S-H gels with prolonged curing. These findings regarding the increased quantity and improved structural orderliness of C-A-S-H and N-A-S-H gels provide a compelling explanation for the age-dependent strength gain in the four PSCM mixtures (Figures 5, 6 and 8).

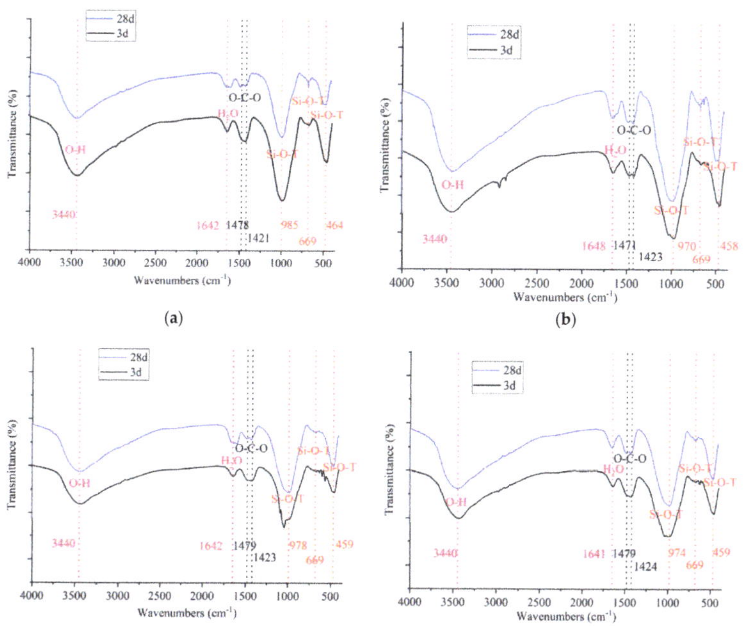

Figure 11. FTIR spectra of the selected PSCM compositions: (**a**) BA0-WG7-KF1, (**b**) BA0-WG7-SS3-KF1, (**c**) BA15-WG7-KF1, and (**d**) BA15-WG7-SS3-KF1.

The bands at 1642 cm^{-1} and 3440 cm^{-1} are assigned to the bending and stretching vibrations of the chemically bonded H–O–H in the hydration products of the samples [65,69,73], which may be imparted by the presence of C-A-S-H and N-A-S-H gels or monosulfate. Additionally, the spectra of all samples exhibit characteristic bands at approximately 1423 cm^{-1} and 1479 cm^{-1}. These bands are attributed to the asymmetric stretching vibrations of O–C–O bonds in the CO_3^{2-} ions present in different polymorphs of $CaCO_3$ (calcite and vaterite) [65–68,70–73]. These polymorphs are likely formed as a result of the carbonation of hydration products during the preparation of powder samples for FTIR analysis. Nevertheless, calcium carbonate crystals were not detectable by XRD in these samples (Figure 10), which may indicate either low crystallinity or a small amount.

3.4.3. TG-DTG Analysis

Thermogravimetric analysis was performed to examine the hydration products of the selected PSCM compositions. The thermogravimetric (TG) and derivative thermogravimetric (DTG) curves for these compositions after curing for 3 days and 28 days are presented in Figure 12.

The result of thermogravimetric analyses of alkali-activated slag and Portland cement samples are discussed in References [71,75–78]. In this paper, the broad endothermic mass loss peak observed below 200 °C in the DTG curve of PSCM paste samples can be attributed to water adsorbed into or bound to the binder to varying degrees [71,77]. Specifically, mass loss peaks at approximately 90 °C are attributed to the removal of free water present in the pore network of the binder or adsorbed from the atmosphere, while mass loss peaks (or shoulders) at about 120 °C are attributed to the dehydration of hydration products, including monosulfate phases (AFm) and C-A-S-H and N-A-S-H gels. The intensity of this broad mass loss peak increases with prolonged curing time, indicating that the ongoing hydration reaction of PSCM results in the formation of more hydration products, such as C-A-S-H and N-A-S-H gels and AFm. This finding is consistent with the FTIR analysis results (Figure 11).

Moreover, the mass loss peaks at approximately 365 °C in the DTG curves of compositions BA0-WG7-KF1 and BA0-WG7-SS3-KF1 cured for 28 days can be attributed to the dehydration of C-A-S-H and N-A-S-H gels [72,78]. In contrast, the mass loss peaks at about 425 °C in the DTG curves of samples BA0-WG7-KF1, BA15-WG7-KF1, and BA15-WG7-SS3-KF1 cured for 3 days are attributed to the dehydroxylation of portlandite, which originates from the hydration of lime. Notably, the absence of a mass loss peak due to portlandite in the DTG curves of the three afore-mentioned compositions cured for 28 days, as well as mixture BA0-WG7-SS3-KF1 cured for both 3 days and 28 days, is evident. This absence is likely due to the consumption of portlandite during the continuous hydration of PSCMs and the carbonation of portlandite during the preparation of powder samples for TG analysis. Additionally, the DTG peaks at approximately 780 °C in all curves of the PSCM samples cured for 3 days and 28 days are attributed to the decarbonization of calcium carbonates.

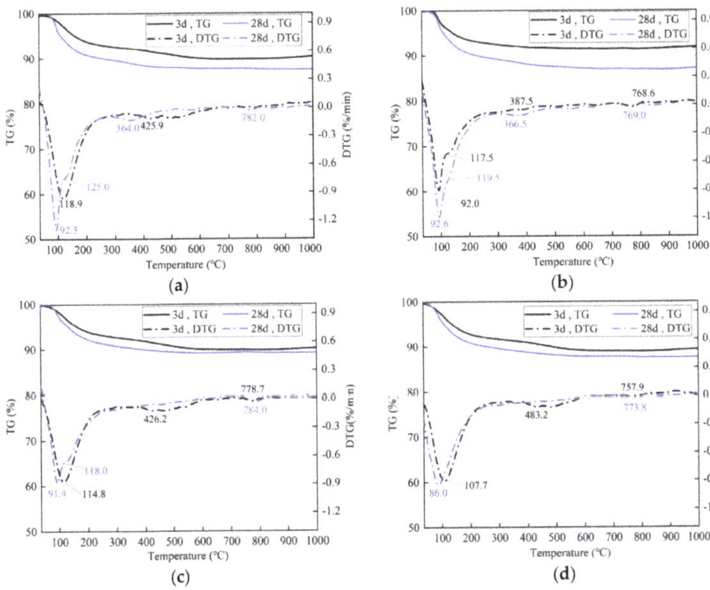

Figure 12. TG-DTG curves of the selected PSCM compositions cured for 3 days and 28 days: (**a**) BA0-WG7-KF1, (**b**) BA0-WG7-SS3-KF1, (**c**) BA15-WG7-KF1, and (**d**) BA15-WG7-SS3-KF1.

3.4.4. SEM-EDS Analysis

The morphologies of the hydration products and microstructures of selected PSCM paste samples cured for 3 days and 28 days were investigated using SEM. The corresponding SEM images are shown in Figure 13.

As illustrated in Figure 13a,d,g,j, the 3-day PSCM samples exhibit a microstructure characterized by cracks, pores, and unhydrated cementitious material particles wrapped in C-A-S-H and N-A-S-H gels. No portlandite (CH) crystals were detected in the SEM images of these samples, likely due to the low CH content at this early stage, which renders them undetectable. By 28 days, the hardened PSCM paste showed a significant reduction in cracks and pores, resulting in a more compact microstructure. This improvement can be attributed to the ongoing hydration as the curing period extended, leading to the substantial consumption of PS, BA, and CH, and the formation of additional C-A-S-H and N-A-S-H, as confirmed by FTIR and TG-DTG analyses. The resulting hydration products reduce porosity, refine the microstructure, and enhance the mechanical properties of the hardened paste. Additionally, no calcium aluminate hydrate crystals were observed in any PSCM samples, including those incorporating sodium sulfate (e.g., BA0-WG7-SS3-KF1 and BA15-WG7-SS3-KF1), where no ettringite or gypsum crystals were found. These findings are consistent with those of XRD, FTIR, and TG analyses.

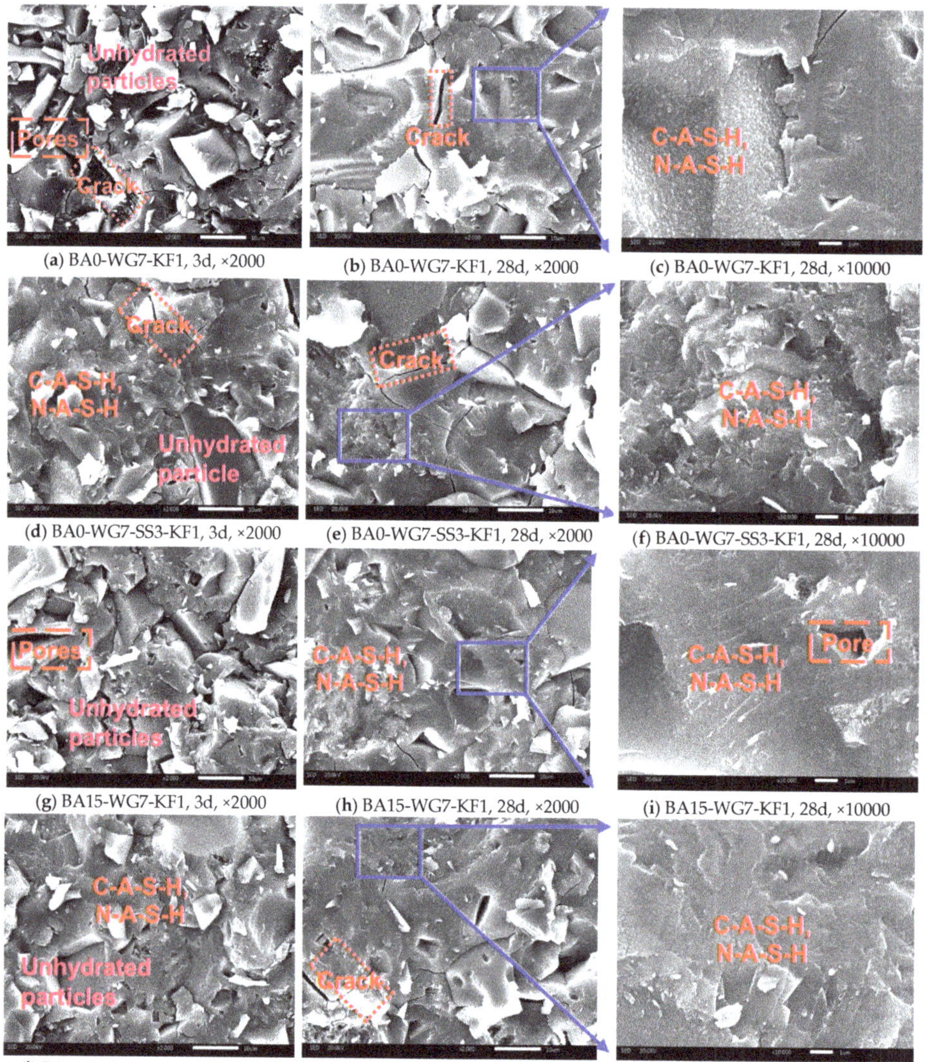

(**a**) BA0-WG7-KF1, 3d, ×2000 (**b**) BA0-WG7-KF1, 28d, ×2000 (**c**) BA0-WG7-KF1, 28d, ×10000

(**d**) BA0-WG7-SS3-KF1, 3d, ×2000 (**e**) BA0-WG7-SS3-KF1, 28d, ×2000 (**f**) BA0-WG7-SS3-KF1, 28d, ×10000

(**g**) BA15-WG7-KF1, 3d, ×2000 (**h**) BA15-WG7-KF1, 28d, ×2000 (**i**) BA15-WG7-KF1, 28d, ×10000

(**j**) BA15-WG7-SS3-KF1, 3d, ×2000 (**k**) BA15-WG7-SS3-KF1, 28d, ×2000 (**l**) BA15-WG7-SS3-KF1, 28d, ×10000

Figure 13. SEM images of selected PSCM paste specimens at different ages.

Figure 14 presents the elemental distribution within the hydration samples of the selected PSCM at 7 days of age, as determined by SEM-EDS analyses.

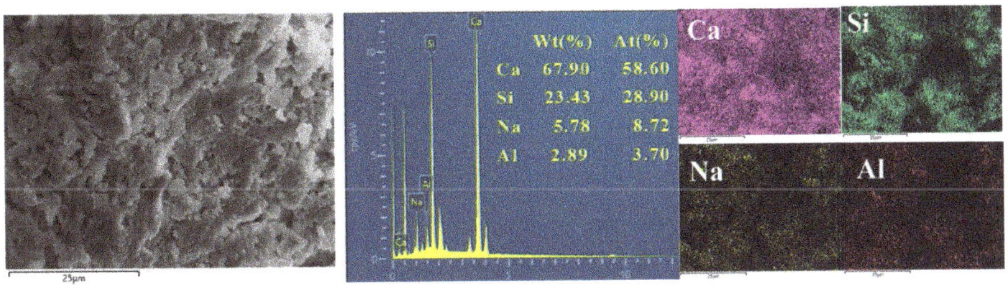

(**a**) BA0-WG7-KF1, 7d, ×2000

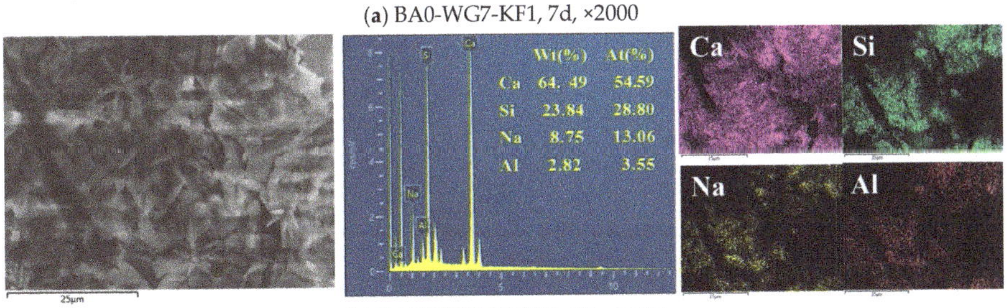

(**b**) BA0-WG7-SS3-KF1, 7d, ×2000

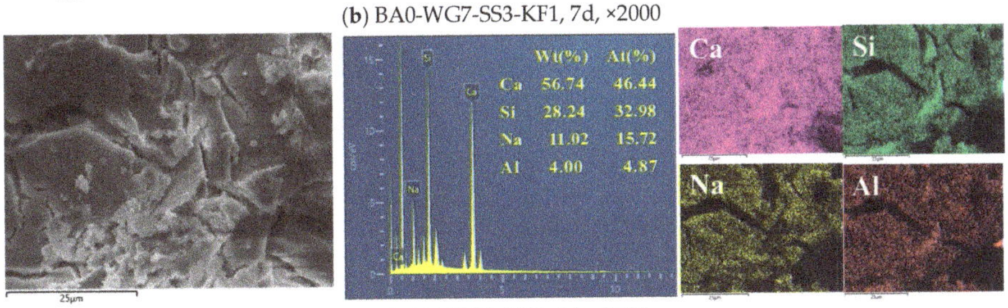

(**c**) BA15-WG7-KF1, 7d, ×2000

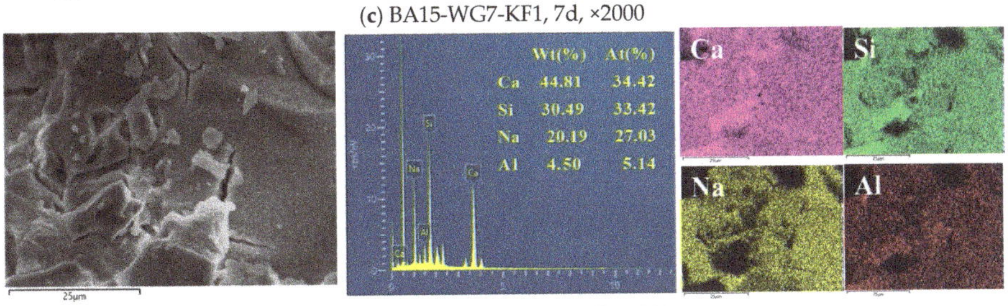

(**d**) BA15-WG7-SS3-KF1, 7d, ×2000

Figure 14. SEM-EDS analysis results for the hydration samples of the selected PSCM at 7 days of age.

As illustrated in Figure 14, the hydration samples of the selected PSCM exhibited high concentrations of calcium (Ca) and silicon (Si), along with trace amounts of aluminum (Al) and sodium (Na). Based on the results presented in the figure, the molar ratios of calcium, aluminum, and sodium to silicon in each hydration sample of the selected PSCMs were calculated. The detailed data are systematically summarized in Table 6.

Table 6. Molar ratios of calcium, aluminum, and sodium to silicon in the hydrated samples of the selected PSCMs.

The Selected PSCM \ Molar Ratios	Ca/Si	Na/Si	Al/Si
BA0-WG7-KF1	2.03	0.30	0.13
BA0-WG7-SS3-KF1	1.90	0.45	0.12
BA15-WG7-KF1	1.41	0.48	0.15
BA15-WG7-SS3-KF1	1.03	0.81	0.15

As shown in Table 6, the incorporation of SS resulted in a decrease in the calcium-to-silicon molar ratio (Ca/Si) from 2.03 to 1.90 (samples BA0-WG7-KF1 and BA0-WG7-SS3-KF1). Conversely, the sodium-to-silicon molar ratio (Na/Si) increased from 0.30 to 0.45 (samples BA0-WG7-KF1 and BA0-WG7-SS3-KF1). Similarly, for the BA15-WG7-KF1 sample, the incorporation of SS led to a reduction in the Ca/Si molar ratio from 1.41 to 1.03 (samples BA15-WG7-KF1 and BA15-WG7-SS3-KF1), as well as an increase in the Na/Si molar ratio from 0.48 to 0.81 (samples BA15-WG7-KF1 and BA15-WG7-SS3-KF1). These findings confirm that the incorporation of SS leads to a decrease in the Ca/Si ratio and an elevation in the Na/Si ratio within the hydration products of PSCMs.

It can also be seen in Table 6 that the incorporation of BA resulted in a marked decrease in the Ca/Si ratio of the hydration products of PSCMs, while both the Na/Si ratio and the Al/Si ratio exhibited significant increases. Specifically, after introducing BA, the Ca/Si ratio in sample BA0-WG7-KF1 decreased from 2.03 to 1.41 (sample BA15-WG7-KF1), while the Na/Si ratio increased from 0.30 to 0.48 and the Al/Si ratio increased from 0.13 to 0.15 (samples BA0-WG7-KF1 and BA15-WG7-KF1), respectively. Similarly, following the addition of BA, the Ca/Si molar ratio in sample BA0-WG7-SS3-KF1 decreased from 1.90 to 1.03 (sample BA15-WG7-SS3-KF1), whereas the Na/Si ratio increased from 0.45 to 0.81 and the Al/Si ratio increased from 0.12 to 0.15 (samples BA0-WG7-SS3-KF1 and BA15-WG7-SS3-KF1), respectively. These findings indicate that the incorporation of BA effectively promotes a reduction in the Ca/Si molar ratio, as well as increases in both the Na/Si and Al/Si ratios within the hydration products of PSCMs.

The SEM-EDS analysis results demonstrate that the hydrated samples of the optimally formulated PSCMs exhibit a reduced Ca/Si ratio, along with increased Al/Si and Na/Si ratios. These observations may provide a plausible explanation for the improved mechanical properties and durability of PSCMs.

Based on the aforementioned microstructural analysis results, it is reasonable to infer that the predominant hydration products of the developed PSCMs are C-A-S-H and N-A-S-H gels, with no detectable crystalline phases such as ettringite, gypsum, or calcium aluminate hydrates present. The absence of these crystalline phases significantly contributes to the material's pronounced shrinkage and the prevalence of microcracks in its microstructure. Future research will focus on an in-depth investigation of the mechanisms governing PSCM shrinkage control.

3.5. Cost Analysis for the Optimal Formulation of PSCMs

Based on the market prices of each component [79], a cost analysis was performed for the optimal PSCM formula (BA15-WG7-SS3-KF1), with the results presented in detail in Table 7.

As shown in Table 6, the total cost of the optimal PSCM formulation is approximately CNY 385 per ton, which is substantially lower than that of Grade 42.5 ordinary Portland cement (CNY 440 per ton). This highlights the significant cost advantages of PSCMs when used in practical applications.

Table 7. Cost analysis of the optimal PSCM formulation developed in this study.

Item \ Components	PSCM						P·O 42.5
	PS	BA	L	WG (by Na_2O)	SS	KF	
Mass percentage, %	80	15	5	13.6 [a]	3.0	1.0	100
Component price [79], CNY/ton	160	100	400	550	860	12,000	440
Component Cost, CNY	128	15	20	74.6	25.8	120	440
Total cost, CNY/ton	383.4						440

[a] The mass percentage of WG is calculated based on its Na_2O mass fraction and the optimal Na_2O dosage specified in the PSCM formulation.

4. Conclusions

A type of clinker-free cementitious material, namely, phosphorus slag-based cementitious material (PSCM), was prepared through alkali activation using industrial solid waste, including granulated electric furnace phosphorus slag (PS) and fluidized bed combustion bottom ash (BA), along with lime (L) as a precursor component. Initially, the effects of varying dosages of alkaline activators, including water glass and sodium sulfate, as well as retarder (potassium fluoride, KF), on the properties (setting time and strength) of PSCM were investigated to determine the optimal formulation. Subsequently, the sulfate resistance and freeze–thaw resistance of PSCM concrete prepared using this optimal formulation were evaluated. Finally, the hydration and microstructure of selected PSCM compositions were characterized using X-ray diffraction (XRD), Fourier transform infrared spectroscopy (FTIR), thermogravimetry (TG), and scanning electron microscope (SEM). The following conclusions have been drawn:

(a) When the mass ratio of precursor components for PSCM, i.e., granulated electric furnace phosphorus slag (PS), fluidized bed combustion bottom ash (BA), and lime (L), is 80:15:5, the optimal dosages of water glass (WG), sodium sulfate (SS), and retarder (KF) are 7%, 3%, and 1%, respectively. The designed PSCM exhibits an initial setting time of 50 min and a final setting time of 70 min, with flexural strengths of 3.5 MPa at 3 days and 7.5 MPa at 28 days, and compressive strengths of 19.0 MPa at 3 days and 64.1 MPa at 28 days. These setting times and strength values comply with the requirements for P.PS 42.5-grade Phosphorous Slag Portland Cement specified in the Chinese standard JC/T 740-2006 [25];

(b) The durability evaluation results indicate that PSCM concrete prepared using the optimized PSCM formulation exhibited superior sulfate resistance and freeze–thaw resistance compared to ordinary Portland cement (OPC) concrete of the same strength grade. Under the test conditions employed in this study, the sulfate resistance grades of PSCM-C30 and PSCM-C50 both exceeded KS90, whereas those of OPC-C30 and OPC-C50 were KS30 and KS60, respectively. Additionally, the freeze–thaw resistance grades of PSCM-C30 and OPC-C30 were F100 and F50, respectively, while those of PSCM-C50 and OPC-C50 were F175 and F150, respectively

(c) The microstructural investigation demonstrates that the primary hydration products of the designed PSCM are calcium aluminosilicate hydrate (C-A-S-H) and sodium aluminosilicate hydrate (N-A-S-H) gels, with no detectable crystalline phases such as ettringite, gypsum, or calcium aluminate hydrates. FTIR and TG-DTG analyses confirm that the hydration reaction of PSCM continues to progress as the curing time increases, leading to the formation of additional C-A-S-H and N-A-S-H gels. SEM observations reveal that the microstructure of the hardened PSCM paste becomes progressively denser with extended curing time. This can be attributed to the ongoing hydration reaction of PSCM, which results in the formation of additional hydration products, thereby significantly improving the microstructure of the hardened paste, and enhancing both its mechanical strength and its durability.

Author Contributions: Conceptualization, Y.P.; methodology, Y.P.; validation, H.L. and J.X.; formal analysis, H.L. and H.Y.; investigation, H.L. and J.X.; resources, G.X.; writing—original draft preparation, H.L.; writing—review and editing, Y.P.; visualization, H.Y.; supervision, Y.P.; project administration, Y.P. and G.X.; funding acquisition, Y.P. and G.X. All authors have read and agreed to the published version of the manuscript.

Funding: This research was funded by the Natural Science Foundation of Hubei Province of China (grant number 2018CFB642), the Hubei Three Gorges Laboratory Opening & Innovation Fund project (grant number SC211006), the Program of Introducing Talents of Discipline to Universities (i.e., the 111 Project of China, grant number D20015), and the 111 Project of Hubei Province (grant number 2021EJD026).

Institutional Review Board Statement: Not applicable.

Informed Consent Statement: Not applicable.

Data Availability Statement: The original contributions presented in this study are included in the article. Further inquiries can be directed to the corresponding author.

Conflicts of Interest: Author Ji Xiao was employed by the company Second Harbor Engineering Company Ltd. The remaining authors declare that the research was conducted in the absence of any commercial or financial relationships that could be construed as a potential conflict of interest.

References

1. He, Y.; Li, Z.; Liu, P. A co-production system of cement and methanol: Unveiling its advancements and potential. *J. Clean. Prod.* **2024**, *473*, 143523. [CrossRef]
2. IEA Cement [WWW Document]. Available online: https://www.iea.org/energy-system/industry/cement (accessed on 26 May 2023).
3. Pitre, V.; La, H.; Bergerson, J.A. Impacts of alternative fuel combustion in cement manufacturing: Life cycle greenhouse gas, biogenic carbon, and criteria air contaminant emissions. *J. Clean. Prod.* **2024**, *475*, 143717. [CrossRef]
4. Van Oss, H.G.; Padovani, A.C. Cement manufacture and the environment part II: Environmental challenges and opportunities. *J. Ind. Ecol.* **2023**, *7*, 93–126. [CrossRef]
5. Cement Sustainability Initiative. Guidelines for the Selection and Use of Fuels and Raw Materials in the Cement Manufacturing Process. World Business Council for Sustainable Development. December 2005. pp. 1–38. Available online: www.wbcsd.org (accessed on 19 November 2019).
6. Shubbar, A.A.; Jafer, H.; Dulaimi, A.; Hashim, K.; Atherton, W.; Sadique, M. The development of a low carbon binder produced from the ternary blending of cement, ground granulated blast furnace slag and high calcium fly ash. *Constr. Build. Mater.* **2018**, *187*, 1051–1060. [CrossRef]
7. Tang, J.; Zhou, W.; Cai, X.; Zhang, D. Feasibility of upcycling red mud for low-CO_2 cement via calcination: A comparative study with FA and GGBFS based in China. *J. Clean. Prod.* **2024**, *476*, 143734. [CrossRef]
8. Liu, J.; Hao, B.; Yu, J.; Xu, Y.; Cao, Y. Research progress on clinker-free alkali-activated binding materials. *Mater. Rep.* **2014**, *1*, 351–354.
9. Sun, Z.; Lin, X.; Liu, P.; Wang, D.; Vollpracht, A.; Oeser, M. Study of alkali-activated slag as alternative pavement binder. *Constr. Build. Mater.* **2018**, *186*, 626–634. [CrossRef]
10. Gong, C.; Yang, N. Effect of phosphate on the hydration of alkali-activated red mud-slag cementitious material. *Cem. Concr. Res.* **2000**, *30*, 1013–1016. [CrossRef]
11. Smirnova, O.; Kazanskaya, L.; Koplík, J.; Tan, H.; Gu, X. Concrete based on clinker-free cement: Selecting the functional unit for environmental assessment. *Sustainability* **2021**, *13*, 135. [CrossRef]

12. Liu, Z.; Ni, W.; Li, Y.; Ba, H.; Li, N.; Ju, Y.; Zhao, B.; Jia, G.; Hu, W. The mechanism of hydration reaction of granulated blast furnace slag-steel slag-refining slag-desulfurization gypsum-based clinker-free cementitious materials. *J. Build. Eng.* **2021**, *44*, 103289. [CrossRef]
13. Wang, K.; Surendra, P.S.; Alexander, M. Effects of curing temperature and NaOH addition on hydration and strength development of clinker-free CKD-fly ash binders. *Cem. Concr. Res.* **2004**, *34*, 299–309. [CrossRef]
14. Zhang, M.; Ma, Q.; Guo, R.; Shi, T.; Huang, L.; Lin, Z.; Yan, F. Utilization of phosphorus slag in cement concrete. *Bull. Chin. Ceram. Soc.* **2019**, *38*, 2464.
15. Lin, S.; Rao, F.; Zheng, Y.; Li, J. Research progress on resource utilization of phosphorus tailings, phosphogypsum and yellow phosphorus slag by geological polymerization. *Conserv. Util. Miner. Resour.* **2021**, *41*, 150–156.
16. Peng, Y.; Zhang, J.; Liu, J.; Ke, J.; Wang, F. Properties and microstructure of reactive powder concrete having a high content of phosphorus slag powder and silica fume. *Constr. Build. Mater.* **2015**, *101*, 482–487. [CrossRef]
17. Chen, X.; Fang, K.; Yang, H.; Peng, H. Hydration kinetics of phosphorus slag cement paste. *J. Wuhan Univ. Technol.-Mater. Sci. Ed.* **2011**, *26*, 142–146. [CrossRef]
18. Zhang, Z.; Wang, Q.; Yang, J. Hydration mechanisms of composite binders containing phosphorus slag at different temperatures. *Constr. Build. Mater.* **2017**, *147*, 720–732. [CrossRef]
19. Allahverdi, A.; Mahinroosta, M. Mechanical activation of chemically activated high phosphorus slag content cement. *Powder Technol.* **2013**, *245*, 182–188. [CrossRef]
20. Yong, C.L.; Mo, K.H.; Koting, S. Phosphorus slag in supplementary cementitious and alkali activated materials: A review on activation methods. *Constr. Build. Mater.* **2022**, *352*, 129028. [CrossRef]
21. Allahverdi, A.; Pilehvar, S.; Mahinroosta, M. Influence of curing conditions on the mechanical and physical properties of chemically-activated phosphorus slag cement. *Powder Technol.* **2016**, *288*, 132–139. [CrossRef]
22. Yang, J.; Huang, J.; He, X.; Su, Y.; Oh, S.K. Shrinkage properties and microstructure of high volume ultrafine phosphorus slag blended cement mortars with superabsorbent polymer. *J. Build. Eng.* **2020**, *29*, 101121. [CrossRef]
23. Shi, C.; Li, Y. Investigation on some factors affecting the characteristics of alkali-phosphorus slag cement. *Cem. Concr. Res.* **1989**, *19*, 527–533.
24. Allahverdi, A.; Abadi, M.; Hossain, K.; Lachemi, M. Resistance of chemically-activated high phosphorus slag content cement against freeze-thaw cycles. *Cold Reg. Sci. Technol.* **2014**, *103*, 107–114. [CrossRef]
25. *JC/T 740-2006*; Portland Phosphorous Slag Cement. China Construction Science & Technology Press: Beijing, China, 2011.
26. Mehdizadeh, H.; Kani, E.N. Rheology and apparent activation energy of alkali-activated phosphorus slag. *Constr. Build. Mater.* **2018**, *171*, 197–204. [CrossRef]
27. Wang, Q.; Huang, Z.; Wang, D. Influence of high-volume electric furnace nickel slag and phosphorus slag on the properties of massive concrete. *J. Therm. Anal. Calorim.* **2017**, *131*, 873–885. [CrossRef]
28. Peng, Y.; Xiao, J.; Zhan, B.; Wei, L.; Xu, G.; Zhang, J. Preparation of phosphorus slag-based rapid repairing materials and properties of its concrete. *Concrete* **2017**, *11*, 173–177.
29. Kumar, S.; Kumar, R.; Bandopadhyay, A.; Alex, T.C.; Kumar, B.R.; Das, S.K.; Mehrotra, S.P. Mechanical activation of granulated blast furnace slag and its effect on the properties and structure of Portland slag cement. *Cem. Concr. Compos.* **2008**, *30*, 679–685. [CrossRef]
30. Li, D.; Wu, X.; Shen, J.; Wang, Y. The influence of compound admixtures on the properties of high-content slag cement. *Cem. Concr. Res.* **2000**, *30*, 45–50. [CrossRef]
31. Xie, F.; Liu, Z.; Zhang, D.; Wang, J.; Wang, D.; Ni, J. The effect of NaOH content on rheological properties, microstructures and interfacial characteristic of alkali activated phosphorus slag fresh pastes. *Constr. Build. Mater.* **2020**, *252*, 119132. [CrossRef]
32. Tang, J.; Min, D.; Wang, A.; Xie, L. Influence of the cement fineness on strengths of cement pastes containing high phosphorus slag. *J. Mater. Civ. Eng.* **2015**, *27*, 04015047.
33. Sun, B.; Ye, G.; De Schutter, G. A review: Reaction mechanism and strength of slag and fly ash-based alkali-activated materials. *Constr. Build. Mater.* **2022**, *326*, 126843.
34. Palacios, M.; Alonso, M.M.; Varga, C.; Puertas, F. Influence of the alkaline solution and temperature on the rheology and reactivity of alkali-activated fly ash pastes. *Cem. Concr. Compos.* **2019**, *95*, 277–284. [CrossRef]
35. Wianglor, K.; Sinthupinyo, S.; Piyaworapaiboon, M.; Chaipanich, A. Effect of alkali-activated metakaolin cement on compressive strength of mortars. *Appl. Clay Sci.* **2017**, *141*, 272–279. [CrossRef]
36. Deng, X.; Guo, H.; Tan, H.; Nie, K.; He, X.; Yang, J.; Wang, Y.; Zhang, J. Effect of organic alkali on hydration of GGBS-FA blended cementitious material activated by sodium carbonate. *Ceram. Int.* **2022**, *48*, 1611–1621. [CrossRef]
37. Yang, R.; Yu, R.; Shui, Z.; Gao, X.; Xie, X.; Zhang, X.; Wang, Y.; He, Y. Low carbon design of an ultra-high performance concrete (UHPC) incorporating phosphorus slag. *J. Clean. Prod.* **2019**, *240*, 118157. [CrossRef]

38. Hu, J. Comparison between the effects of superfine steel slag and superfine phosphorus slag on the long-term performances and durability of concrete. *J. Therm. Anal. Calorim.* **2017**, *128*, 1251–1263. [CrossRef]
39. Li, D.; Shen, J.; Mao, L.; Wu, X. The influence of admixtures on the properties of phosphorus slag cement. *Cem. Concr. Res.* **2000**, *30*, 1169–1173. [CrossRef]
40. Gao, P.; Lu, X.; Yang, C.; Li, X.; Shi, N.; Jin, S. Microstructure and pore structure of concrete mixed with superfine phosphorus slag and superplasticizer. *Constr. Build. Mater.* **2008**, *22*, 837–840. [CrossRef]
41. Shi, C.; Qu, B.; Provis, J.L. Recent progress in low-carbon binders. *Cem. Concr. Res.* **2019**, *122*, 227–250. [CrossRef]
42. Liu, D. Effect of phosphorus slag and fly ash on anti-corrosion property of high belite cement. *Appl. Mech. Mater.* **2012**, *117–119*, 1231–1234. [CrossRef]
43. Soleimani, M.A.; Naghizadeh, R.; Mirhabibi, A.; Golestanifard, F. The influence of phosphorus slag addition on microstructure and mechanical properties of metakaolin-based geopolymer pastes. *Ceram.-Silik.* **2013**, *57*, 33–38.
44. Vafaei, M.; Allahverdi, A.; Dong, P.; Bassim, N.; Mahinroosta, M. Resistance of red clay brick waste/phosphorus slag-based geopolymer mortar to acid solutions of mild concentration. *J. Build. Eng.* **2021**, *34*, 102066. [CrossRef]
45. Shi, C.; He, F.; Angel, P. Classification and characteristics of alkali-activated cement. *J. Chin. Ceram. Soc.* **2012**, *40*, 69–75.
46. Shi, C.; Li, Y.; Tang, X. A preliminary study on the mechanism of activation of phosphorus slag. *J. Southeast Univ. (Nat. Sci. Ed.)* **1989**, *19*, 141–145.
47. Mehdizadeh, H.; Kani, E.N.; Sanchez, A.P.; Fernandez-Jimenez, A. Rheology of activated phosphorus slag with lime and alkaline salts. *Cem. Concr. Res.* **2018**, *113*, 121–129. [CrossRef]
48. Jia, R.; Wang, Q.; Luo, T. Understanding the workability of alkali-activated phosphorus slag pastes: Effects of alkali dose and silicate modulus on early-age hydration reactions. *Cem. Concr. Compos.* **2022**, *133*, 104649. [CrossRef]
49. Jia, R.; Wang, Q.; Luo, T. Mechanisms and differences between sodium and magnesium sulfate attacks on alkali-activated phosphorus slag. *Constr. Build. Mater.* **2023**, *403*, 133117. [CrossRef]
50. Huang, Z.; Deng, L.; Che, D. Development and technical progress in large-scale circulating fluidized bed boiler in China. *Front. Energy* **2020**, *14*, 699–714. [CrossRef]
51. *GB/T 17671-2021*; Test Method of Cement Mortar Strength (ISO Method). Standards Press of China: Beijing, China, 2021.
52. *GB/T 1346-2024*; Test Methods for Water Requirement of Standard Consistency, Setting Time and Soundness of the Portland Cement. Standards Press of China: Beijing, China, 2024.
53. *GB/T 50081-2019*; Standard for Test Methods of Concrete Physical and Mechanical Properties. China Architecture & Building Press: Beijing, China, 2019.
54. *GB/T 50082-2024*; Standard for Test Methods of Long-Term Performance and Durability of Ordinary Concrete. China Architecture & Building Press: Beijing, China, 2024.
55. Yang, N. Physical and chemical basis of alkali-activated materials formation (I). *J. Chin. Ceram. Soc.* **1996**, *24*, 209–215.
56. Yang, N. Physical and chemical basis of alkali-activated materials formation (II). *J. Chin. Ceram. Soc.* **1996**, *24*, 459–465.
57. Cong, P.; Du, R.; Gao, H.; Chen, Z. Comparison and assessment of carbon dioxide emissions between alkali-activated materials and OPC cement concrete. *J. Traffic Transp. Eng. (Engl. Ed.)* **2024**, *11*, 918–938. [CrossRef]
58. Cong, P.; Cheng, Y.; Ge, W.; Zhang, A. Mechanical, microstructure and reaction process of calcium carbide slag-waste red brick powder based alkali-activated materials (CWAAMs). *J. Clean. Prod.* **2022**, *331*, 129845. [CrossRef]
59. Duxson, P.; Provis, J.L.; Lukey, G.C.; Van Deventer, J.S. The role of inorganic polymer technology in the development of 'green concrete'. *Cem. Concr. Res.* **2007**, *37*, 1590–1597. [CrossRef]
60. Palomo, A.; Grutzeck, M.W.; Blanco, M.T. Alkali-activated fly ashes: A cement for the future. *Cem. Concr. Res.* **1999**, *29*, 1323–1329. [CrossRef]
61. Qin, X.; Wang, P.; Liu, L.; Wang, M.; Xin, J. Sensitivity analysis of microstructure parameters and mechanical strength during consolidation of cemented paste backfill. *Math. Probl. Eng.* **2018**, *2018*, 5170721. [CrossRef]
62. Lei, D.; Guo, L.; Sun, W.; Liu, J.; Miao, C. Study on properties of untreated FGD gypsum-based high-strength building materials. *Constr. Build. Mater.* **2017**, *153*, 765–773. [CrossRef]
63. Feng, Y.; Xue, Z.; Zhang, B.; Xie, J.; Chen, C.; Tan, J.; Zhao, C. Effects of phosphogypsum substitution on the performance of ground granulated blast furnace slag/fly ash-based alkali-activated binders. *J. Build. Eng.* **2023**, *70*, 106387. [CrossRef]
64. Kurdowski, W. *Cement and Concrete Chemistry*; Springer Science: Dordrecht, The Netherlands, 2014.
65. Yang, N.; Yue, W. *The Handbook of Inorganic Metalloid Materials Atlas*; Wuhan University of Technology Press: Wuhan, China, 2000.
66. Li, N.; Farzadnia, N.; Shi, C. Microstructural changes in alkali-activated slag mortars induced by accelerated carbonation. *Cem. Concr. Res.* **2017**, *100*, 214–226. [CrossRef]
67. Shao, Y.; Rostami, V.; He, Z.; Boyd, A.J. Accelerated carbonation of Portland limestone cement. *J. Mater. Civ. Eng.* **2014**, *26*, 117–124. [CrossRef]

68. Legodi, M.; De Waal, D.; Potgieter, J. Quantitative determination of $CaCO_3$ in cement blends by FT-IR. *Appl. Spectrosc.* **2001**, *55*, 361–365. [CrossRef]
69. Peng, Y.; Liu, Y.; Zhan, B.; Xu, G. Preparation of autoclaved aerated concrete by using graphite tailings as an alternative silica source. *Constr. Build. Mater.* **2021**, *267*, 121792. [CrossRef]
70. Bernal, S.A.; de Gutiérrez, R.M.; Pedraza, A.L.; Provis, J.L.; Rodriguez, E.D.; Delvasto, S. Effect of binder content on the performance of alkali-activated slag concretes. *Cem. Concr. Res.* **2011**, *41*, 1–8. [CrossRef]
71. Walkley, B.; Nicolas, R.S.; Sani, M.A.; Rees, G.J.; Hanna, J.V.; van Deventer, J.S.J.; Provis, J.L. Phase evolution of C-(N)-A-S-H/N-A-S-H gel blends investigated via alkali-activation of synthetic calcium aluminosilicate precursors. *Cem. Concr. Res.* **2016**, *89*, 120–135. [CrossRef]
72. Bernal, S.A.; Provis, J.L.; Rose, V. Evolution of binder structure in sodium silicate-activated slag-metakaolin blends. *Cem. Concr. Compos.* **2011**, *33*, 46–54. [CrossRef]
73. Vafaei, M.; Allahverdi, A. Strength development and acid resistance of geopolymer based on waste clay brick powder and phosphorus slag. *Struct. Concr.* **2019**, *20*, 1596–1606. [CrossRef]
74. Palacios, M.; Gismera, S.; Alonso, M.M.; De Lacaillerie, J.D.E.; Lonthenbach, B.; Favier, A.; Brumaud, C.; Puertas, F. Early reactivity of sodium silicate-activated slag pastes and its impact on rheological properties. *Cem. Concr. Res.* **2021**, *140*, 106302. [CrossRef]
75. Peng, Y.; Hu, S.; Ding, Q. Preparation of reactive powder concrete using fly ash and steel slag powder. *J. Wuhan Univ. Technol.-Mater. Sci. Ed.* **2010**, *25*, 349–354. [CrossRef]
76. Bernal, S.A.; Rodríguez, E.D.; Mejía de Gutiérrez, R.; Gordillo, M.; Provis, J.L. Mechanical and thermal characterisation of geopolymers based on silicate-activated metakaolin/slag blends. *J. Mater. Sci.* **2011**, *46*, 5477–5486. [CrossRef]
77. Alarcon-Ruiz, L.; Platret, G.; Massieu, E.; Ehrlacher, A. The use of thermal analysis in assessing the effect of temperature on a cement paste. *Cem. Concr. Res.* **2005**, *35*, 609–613. [CrossRef]
78. Bernal, S.A.; de Gutierrez, R.M.; Provis, J.L.; Rose, V. Effect of silicate modulus and metakaolin incorporation on the carbonation of alkali silicate-activated slags. *Cem. Concr. Res.* **2010**, *40*, 898–907. [CrossRef]
79. Available online: http://zj.yichang.gov.cn/content-62527-985084-1.html (accessed on 4 June 2025).

solids

MDPI

Article

Performance Evaluation of Low-Grade Clay Minerals in LC^3-Based Cementitious Composites

Nosheen Blouch [1], Syed Noman Hussain Kazmi [2], Nijah Akram [3], Muhammad Junaid Saleem [4,5], Imran Ahmad Khan [5,*], Kashif Javed [5], Sajjad Ahmad [2] and Asfandyar Khan [5,6]

1 Department of Building and Architectural Engineering, Faculty of Engineering and Technology, Bahauddin Zakariya University, Multan 60800, Pakistan; nosheenbalouch@bzu.edu.pk
2 Department of Civil Engineering, Faculty of Engineering and Technology, Bahauddin Zakariya University, Multan 60800, Pakistan; kazminoman276@gmail.com (S.N.H.K.); sajjadahmad@bzu.edu.pk (S.A.)
3 Department of Architectural Engineering Technology, Punjab Tianjin University of Technology, Lahore 53720, Pakistan; akram.nijah@ptut.edu.pk
4 Department of Textile and Fashion, Punjab Tianjin University of Technology, Lahore 54000, Pakistan; junaid.saleem@ptut.edu.pk
5 Department of Textile and Apparel Science, University of Management and Technology, Lahore 54770, Pakistan; kashif.javed@umt.edu.pk (K.J.); asfandyar.khan@umt.edu.pk (A.K.)
6 Department of Textile Engineering, Daffodil International University (DIU), Daffodil Smart City, Birulia, Dhaka 1216, Bangladesh
* Correspondence: imran.ahmad@umt.edu.pk; Tel.: +92-313-9374141

Abstract

The cements industry is increasingly under pressure to reduce carbon emissions while maintaining performance standards. Limestone calcined clay cement (LC^3) presents a promising low-carbon alternative; however, its performance depends significantly on the type and reactivity of clay used. This study investigates the effect of three common low-grade clay minerals—kaolinite, montmorillonite, and illite—on the behavior of LC^3 blends. The clays were thermally activated and characterized using X-ray diffraction (XRD), thermogravimetric analysis (TGA), X-ray fluorescence spectroscopy (XRF), and Blaine air permeability testing to evaluate their mineralogical composition, thermal behavior, chemical content, and fineness. Pozzolanic reactivity was assessed using the modified Chapelle test. Microstructural development was examined through scanning electron microscopy (SEM) of the hydrated specimens at 28 days. The results confirmed a strong correlation between clay reactivity and hydration performance. Kaolinite showed the highest reactivity and fineness, contributing to a dense microstructure with reduced portlandite and enhanced formation of calcium silicate hydrate. Montmorillonite demonstrated comparable strength and favorable hydration characteristics, while illite, though less reactive initially, showed acceptable long-term behavior. Although kaolinite delivered the best overall performance, its limited availability and higher cost suggest that montmorillonite and illite represent viable and cost-effective alternatives, particularly in regions where kaolinite is scarce. This study highlights the suitability of regionally available, low-grade clays for use in LC^3 systems, supporting sustainable and economically viable cement production.

Keywords: clay minerals; kaolinite; illite; montmorillonite; reactivity; modified Chapelle tests

1. Introduction

The cement industry is confronted with sustainability issues, which have led to the development of various technologies aimed at reducing CO_2 emissions. LC^3 has emerged as a promising ternary blended composite and has gained significant attention in recent years. Its comparable mechanical properties and favorable environmental impact make it a sustainable alternative to OPC. LC^3 has gained widespread importance as it uses easily sourced materials that require little processing and have a comparatively high rate of substitution for cement clinker. This results in a composite cement with lesser embodied energy and reduced carbon dioxide emissions, analogized to Portland cement.

The usual composition of LC^3 is 30%, 15%, 50%, and 5% of calcined clay, limestone, cement, and gypsum, respectively. The calcined kaolinite has been extensively studied in LC^3 systems and has demonstrated excellent pozzolanic reactivity [1–3]. Its beneficial use in LC^3 is well documented; for instance, replacing 50% of cement mortar with metakaolin and limestone significantly enhances the mechanical performance and durability properties of the cement mortar [4–6]. Recent focus has shifted to evaluating the viability of alternative clay minerals—specifically illite and montmorillonite—as prospective pozzolanic materials in LC^3 systems. Despite their natural abundance and potential for local sourcing, these clays have been less investigated. However, the pozzolanic characteristics of calcined kaolinites, compared to other clay minerals, for example, illite and montmorillonite, in LC^3 systems have not been frequently studied and remain inadequately understood. Due to their use in producing paint, paper, and ceramic goods, kaolinite clays are highly competitive with other industries. The development of LC^3 might be less expensive and more accessible if other commonly available clay minerals, like montmorillonite and illite, were used.

Clays are hydrous aluminum silicate-based minerals with layered sheet structures and particles that are usually smaller than 2 μm [7]. These minerals are essential to fine-grained metamorphic rocks like slate and phyllite and sedimentary rocks like shale, mudstone, and siltstone. Typically, they form through weathering and hydrothermal reactions at low temperatures. The three main clay minerals are kaolinite [$Al_2Si_2O_5(OH)_4$], montmorillonite [$(Na,Ca)_{0.33}(Al,Mg)_2(Si_4O_{10})(OH)_2\cdot nH_2O$], and illite [$(Al_2)(Si_{4-x}Al_x)O_{10}(OH)_2\cdot K_{1-x}$] [8] (Figure 1). These minerals have a layered structure consisting of two distinct types. The layered structure of these minerals is made up of two different kinds of layers. In the first layer, four oxygen atoms form bonds with the Si^{4+} cation in silicon tetrahedra. On the other hand, the second layer consists of an octahedral-shaped aluminum hydroxide layer with an Al^{3+} cation at its center [9].

In some clay minerals, magnesium (Mg^{2+}) can substitute for aluminum (Al^{3+}) in the octahedral layer [10]. Kaolinite is categorized as a 1:1 clay mineral, with H-bonds linking the OH^- groups of one layer to the O^{2-} groups of another, resulting in a stable structure that restricts cation intercalation [9]. Montmorillonite and illite, on the other hand, are 2:1-type clay minerals where the oxygen anions (O^{2-}) from one layer are positioned next to those from an adjacent layer [11].

Heating raw clays produces metaclays, which are the source of calcined clays' reactivity. As pozzolanic substances, these metaclays react with portlandite to produce (1) ettringite, (2) calcium silicate hydrate (C-S-H), and (3) AFm phases. The addition of limestone to Portland cement causes the calcite to react with the clinker's C^3A, forming both hemi and monocarboaluminate phases. In LC^3, the aluminate from metaclays further reacts with calcite, facilitating the development of carboaluminate phases.

Most previous studies on metakaolin's impact on hydration have focused on high- to medium-grade kaolinitic clays [12–15]. However, these materials are not economically viable, particularly in regions like the Southern Hemisphere, where they can be up to three times more expensive than Portland cement [16]. These cost disparities emphasize the economic advantage of using illite and montmorillonite in LC^3 formulations, especially in regions with constrained construction budgets. Although they have lower alumina content, their structural and chemical properties indicate they could still offer satisfactory levels of pozzolanic reactivity if they are calcined correctly. There is limited information on calcined common clays, such as low-grade illite and montmorillonite, which are abundant and have significant potential as clinker substitutes [17–19]. Alongside chemical and strength performance, the influence of supplementary cementitious materials on microstructure and crack propagation is increasingly being explored. The behavior of fractures in cementitious composites is intricately connected to their hydration products, the refinement of pores, and the characteristics of interfacial transition zones. Supplementary cementitious materials like calcined clays have the capacity to modify these microstructural characteristics, which may enhance durability and resistance to cracking. Proposed advanced numerical models simulate this behavior, demonstrating the impact of reactive mineral additions on stress redistribution and crack development under load [20,21]. Integrating these elements offers a more comprehensive understanding of the structural effects associated with the use of alternative clays in LC^3 formulations.

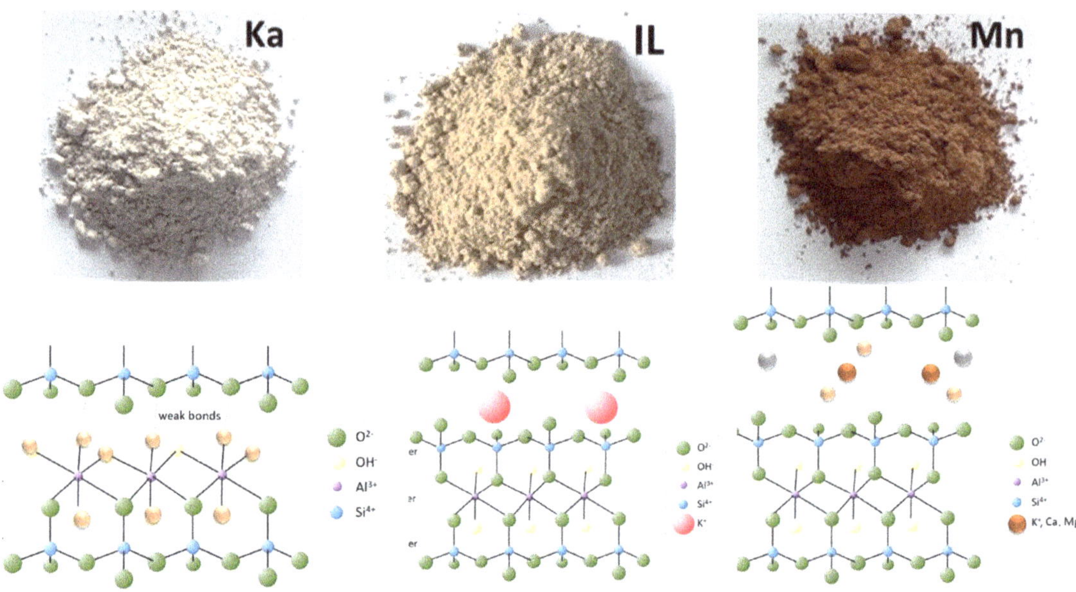

Figure 1. Clay minerals (color and their expected structure based on Li et al. [9].

This research investigates the influence of calcined kaolinite, illite, and montmorillonite on the hydration mechanisms and strength development of LC^3 blends. The activated clays were tested for reactivity using the modified Chapelle test. The hydration mechanisms were analyzed using XRD and TGA, while strength development was evaluated through the heat of hydration, setting time, consistency, and compressive strength evaluation.

2. Materials and Methods

2.1. Materials and Activation

The current work explored the potential use of three low-grade clays, kaolinite (Ka), illite (IL), and montmorillonite (Mn) for LC^3 production from the different sources of

Pakistan. Ka clays were explored from Pai Khel, district Mianwali (Punjab). Mn clay was collected from the Hub, District Lasbela, (Baluchistan), and the IL clay was collected in Jhampir (Sindh), Pakistan. Ordinary Portland Cement (OPC) and limestone used were collected from the Maple Leaf Cement Factory Pvt. Limited, Pakistan.

Kaolinite ($Al_2Si_2O_5(OH)_4$) consists of layers formed by two sheets linked by hydrogen bonds, resulting in a stable non-swelling structure [8]. Illite and montmorillonite are 2:1 clay minerals, having an octahedral sheet of alumina packed between two silicon tetrahedral sheets [8]. However, the montmorillonite has weaker bonds that are easy to break, while the illite layers are firmly bonded by potassium and require a high activation temperature [22]. Therefore, the raw clays were heat treated at two different temperatures. Ka and Mn were calcined at 750 °C, while IL required 850 °C due to its thermal stability. Each 1 kg batch was heated in an alumina crucible in a laboratory furnace for one hour and then rapidly cooled. The calcination temperatures were selected based on the dehydroxylation behavior observed from thermal analysis.

2.2. Characterization Techniques

The mineralogical and chemical compositions of the raw and calcined clays were evaluated using several XRD, XRF, PSD, and TGA. International standard protocols were followed during the testing procedures. X-ray fluorescence spectroscopy was conducted on powdered samples employing a PANalytical Smart Zetium X-ray fluorescence spectrometer. X-ray diffraction (XRD) analysis was carried out using a BTX III Benchtop XRD Analyzer. The samples were ground into a powder and passed through a 200-mesh screen after curing for seven and twenty-eight days. The analysis used Cu Kα radiation at 30 mA and 30 kV. The scanning angle lies between 5° and 50° 2θ, with a step size of 0.025° and a scanning time of 1.58 s per step. TG/DTA analysis was performed using a LINSEIS STA PT 1600 (TG-DSC), heating the samples at a 10 °C/min rate from 50 °C to 950 °C using N_2 as a purge gas. Additionally, a correlation matrix was developed to link the clay's reactivity with the efficiency of LC^3.

Blaine fineness assessments were performed in accordance with ASTM C204 standards via a Blaine air permeability device. Samples were subjected to oven drying, grinding, and sieving through a 75 µm (No. 200) mesh prior to testing to ensure homogeneous packing and eliminate agglomerates. The fineness values were documented in cm^2/g, indicating the particular surface area associated with particle arrangement and possible reactivity in cementitious systems.

Blaine's fineness assessments revealed substantial disparities between raw and calcined clays. Upon calcination, all clays demonstrated a significant increase in specific surface area resulting from structural degradation and dehydroxylation [23]. Calcined kaolinite exhibited the highest fineness among the clays, signifying enhanced grindability and a morphology favorable for achieving a large surface area [24]. Montmorillonite demonstrated a significant enhancement in fineness following calcination, nearly matching that of kaolinite. Illite exhibited the least fineness among the calcined clays, indicative of its denser structure and reduced grinding efficiency [25]. Compared to OPC, all calcined clays exhibited significantly higher surface areas, indicating an improved potential for pozzolanic reaction. Limestone demonstrated the greatest fineness due to its soft and easily grindable characteristics. The variations in fineness are anticipated to impact the early dissolution behavior and, in turn, influence the hydration kinetics of LC^3 systems [26]. The Blaine fineness analysis of clay samples, cement, and limestone are tabulated in Table 1. Figure 2 shows the particle size distribution (PSD) determined using a Malvern Mastersizer 2000 laser diffraction analysis.

Table 1. XRF analysis of clay minerals, limestone, and mortar blends (OPC and LC[3]) in raw (R) and calcined (C) forms.

Clay Type	SiO_2	Al_2O_3	Fe_2O_3	CaO	MgO	K_2O	Na_2O	SO_3	Cl	Blain Finess (cm^2/g)	LO1
Ka (R)	58.7	28.8	3.5	2.3	0.2	0.3	0.1	0.2	0.0	6552	12.15
Ka (C)	53.16	34.28	3.36	2.67	0.04	0.01	0.02	0.06	0.00	10,930	2.97
Mn (R)	41.45	29.23	2.32	9.60	0.68	2.55	0.65	0.03	-	4545	10.34
Mn (C)	41.45	27.13	2.02	8.76	0.60	2.55	0.65	0.03	-	10,000	0.69
IL (R)	56.40	19.57	6.98	4.73	1.71	1.49	1.39	0.1	0.72	4650	10.29
IL (C)	54.06	18.64	7.13	4.64	1.80	1.57	1.59	0.08	0.17	8765	0.48
Limestone	19.28	5.44	2.09	42.5	0.96	0.78	0.37	0.15	0.00	11,535	43.93
OPC	19.4	4.81	3.54	60.9	2.58	2.65	0.25	0.77	0	3200	2.5

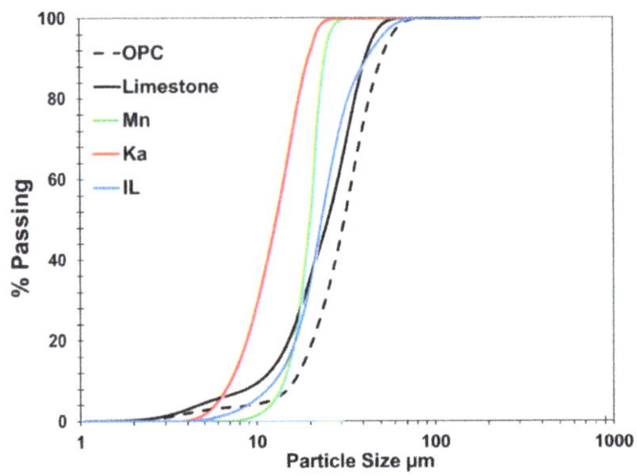

Figure 2. Particle size distribution (PSD) analysis of the raw material used.

The XRD analysis of Ka indicated the presence in kaolinite minerals, with a prominent peak at 11.3° 2θ (Figure 3) [27]. The IL sample exhibited diffraction peaks corresponding to illite, along with minor quantities of quartz, kaolinite, and calcite [28]. The Mn sample showed sharp peaks of montmorillonite near 6° 2θ [29], accompanied by detectable peaks of kaolinite, quartz, and traces of illite and calcite. After thermal activation, the XRD patterns of the calcined clays were analyzed to assess the effectiveness of the calcination process. Kaolinite demonstrated a significant reduction in crystallinity after being calcined at 750 °C, as indicated by the absence of kaolinite peaks, which signifies its transformation into an amorphous phase. Calcined Mn exhibited the removal of montmorillonite peaks and the appearance of weak peaks for albite and calcite [30]. In the calcined IL sample, minor illite peaks remained, indicating partial transformation. Quartz peaks remained stable in all samples, suggesting that quartz was unaffected by calcination.

The thermogram and derivative thermal analysis are revealed in Figure 4a,b. The thermograms revealed variations in mass loss patterns among the clays. Kaolinite (Ka) exhibited minimal mass loss below 300 °C, while illite (IL) showed slight mass loss, and montmorillonite (Mn) displayed the most significant mass reduction in this range [31]. This region corresponds to the removal of adsorbed water. In the 300–600 °C range, all samples showed a second mass loss peak attributed to the dehydroxylation of structural hydroxyl groups. The dehydroxylation of Mn and IL extended up to approximately 750 °C and 800 °C, respectively [32,33]. The DTG curves indicated distinct differences in thermal decomposition behavior, suggesting variations in mineral composition and structural stability among the clays. These thermal trends were consistent with XRD observations regarding phase transformation upon calcination [32]. The XRD and DTG results were consistent with each other.

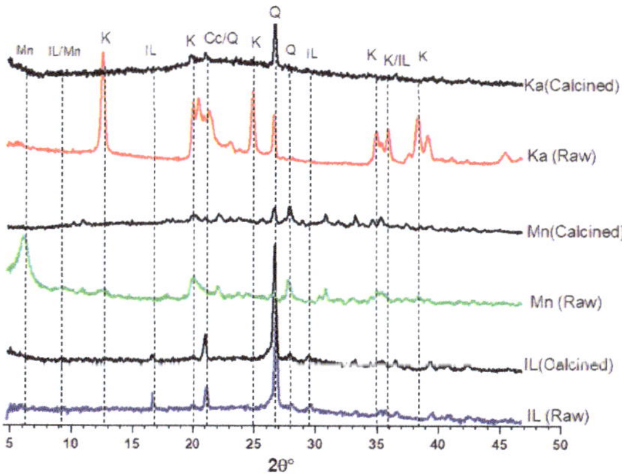

Figure 3. X-ray diffraction analysis of the raw clays. where K is kaolinite, M is montmorillonite, Q is quartz, H is hematite, and I is illite.

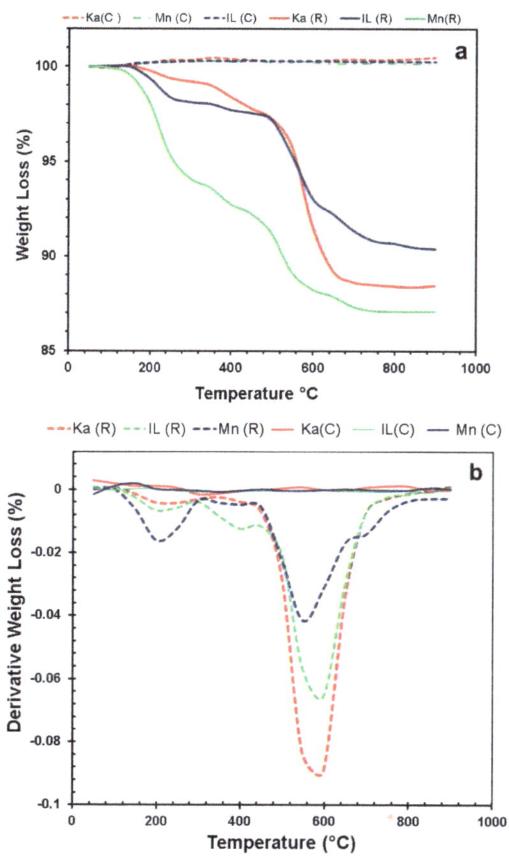

Figure 4. Weight loss (**a**) and derivative weight loss (**b**) of raw and calcined clay minerals.

The activated clays must demonstrate sufficient pozzolanic reactivity to ensure the rigorous integrity of the binder. The modified Chapelle test was employed to determine the reactivity of the calcined clays. This test measures the reactivity of clay by heating a mixture of calcium oxide and calcined clay under reflux conditions. The process involves mixing one gram of calcined clay with two grams of CaO and 250 mL of deionized water. Following this, the blend is heated for sixteen hours at 90 °C. After cooling, 250 mL of deionized water is combined with 60 g of sucrose. The resulting mixture is titrated against HCl (0.1 M). The amount of bound portlandite is calculated based on the volumes of HCl solution added for the blank and calcined clay samples. The pozzolanic activity is calculated using Formula (1). The pictorial procedure of the tests is shown in Figure 5.

$$PA = 2\frac{V_1 - V_2}{V_1} \cdot \frac{74}{56} .1000 \quad (1)$$

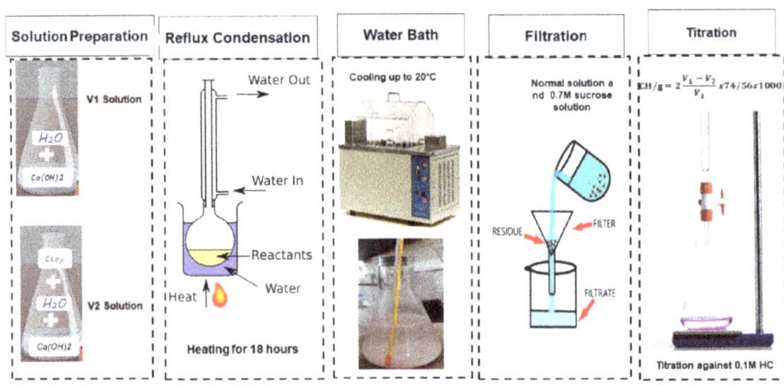

Figure 5. Pictorial presentation of the modified Chapelle test.

2.3. Development of LC^3

To develop LC^3, a mixture of 2 parts clay and 1 part limestone was used to replace 50% of OPC based on previous research [34]. The cement paste was mixed using a mechanical mixer at 1600 rpm for 3 min. The LC^3 mortar was fabricated with a 1:2.75 binder-to-sand ratio and a 0.485 water-to-binder ratio following ASTM specifications. The sand used in the mortar formulation was well-graded Lawrence pure sand, characterized by a fineness modulus of 2.7, a specific gravity of 2.65, water absorption of 1.9%, and a loose density of 1340 kg/m^3. The mortar mixture design is displayed in Table 2, and Figure 6 illustrates it. A Hobart mixer was used to blend the mortar for three minutes. Three layers of mortar were poured into the 50 mm mold, and each layer was tamped 25 times. Finally, a table vibrator was used for a minute to compact it. The molds were wrapped in a low-density polyethylene sheet to prevent heat loss, keep the temperature steady at 25 °C, and maintain relative humidity at 75%. Following 24 h, the cubes were removed from the molds and cured in a humid room until they reached the necessary age.

Table 2. Mix the proportions of OPC LC^3-mortar pastes.

	OPC kg/m^3	Clay (kg/m^3)			Limestone kg/m^3	Sand kg/m^3	Water kg/m^3
		Ka	Mn	IL			
LC^3-Ka	170	212	---	----	98	1039	115
LC^3-Mn	170	----	212	---	98	1039	115
LC^3-IL	170	---	---	212	98	1039	115
Control Mix	569	---	---	---	---	1039	115

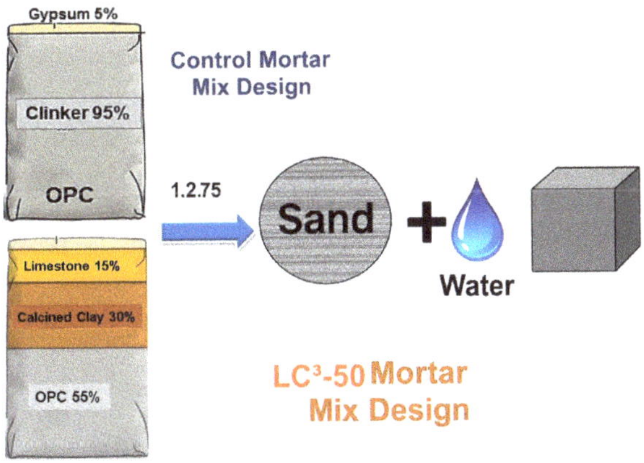

Figure 6. Pictorial representation for the development of LC^3 and OPC blends.

2.4. Experimental Setup

Several tests were performed on the fresh and hardened cement mortar blend. Fresh properties, including normal consistency and setting times, were assessed according to ASTM standards, according to ASTM C187 (2016) and ASTM C230 (2020) [35,36], respectively. The compressive strength of the hardened material was evaluated at 7 and 28 days, with results averaged from three specimens using a universal testing equipment that applies a load of 2000 kN and operates at a loading rate of 1.2 mm/min, in compliance with ASTM C109 standards [35]. An isothermal heat conduction calorimeter was used to measure the heat of hydration, monitoring early-stage hydration for up to seven days following ASTM C1702 [36]. TGA and XRD were performed on powdered, cured samples to evaluate hydration mechanisms and phase development.

Figure 7 depicts the detailed methodology, workflow, and strength development mechanism of LC^3 mortar incorporating kaolinite, montmorillonite, and illite. It includes raw material processing, calcination to form metakaolin, meta-smectite, and meta-illite, followed by reactivity assessment using the modified Chapelle test. LC^3 mortar is developed by mixing calcined clay, limestone, gypsum, and water. The following stages involve hydration product analysis (XRD, TGA), performance testing (strength, flowability, setting time), and heat of hydration monitoring (1–7 days) to evaluate the influence of each clay type.

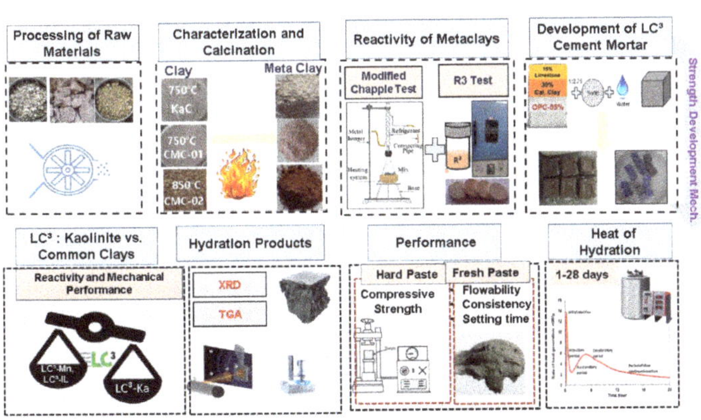

Figure 7. Comprehensive methodology of the performed research work.

3. Results and Discussion

3.1. Reactivity of the Clays

3.1.1. The Modified Chapelle Test

The pozzolanic reactivity of calcined clays was assessed using the modified Chapelle test. Figure 8 illustrates the portlandite consumption by different types of calcined clay minerals, categorized by their crystalline structure. The amount of portlandite reacted varies according to the type of clay, with Ka consuming the most portlandite (1072 mg of Ca $(OH)_2$ per gram of Ka), followed by Mn (881 mg of $Ca(OH)_2$ per gram of Mn), and IL (783 mg of $Ca(OH)_2$ per gram of IL). The Ka clay exhibited the highest reactivity, indicating higher reactivity due to its 1:1 kaolinitic structure [32]. The structural mineralogy of IL clay is the reason for its decreased reactivity [37]. The high testing temperature contributed to significant portlandite consumption across all clays.

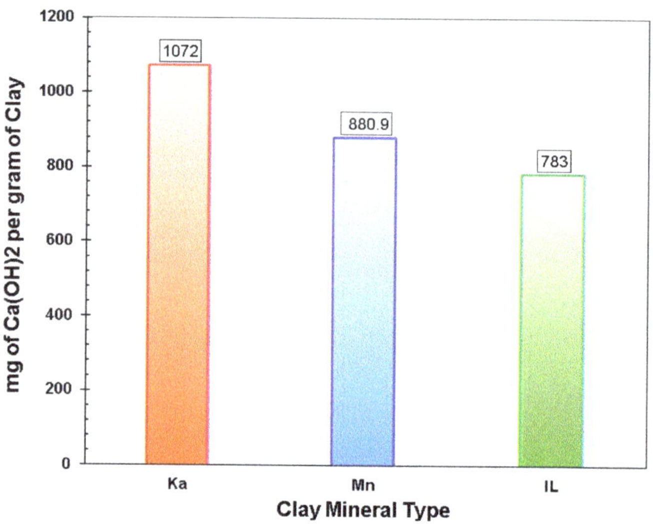

Figure 8. The consumption mg of Ca $(OH)_2$ per gram of clay minerals.

3.1.2. Mineralogical Evaluation of Calcined Clays

To comprehend the reactivity tendencies identified in the Chapelle test, clays were examined for their mineralogical, chemical, and physical properties. Table 3 summarizes the reactivity-related characteristics of the three clays. Ka, a 1:1 layered clay, demonstrates the most advantageous reactivity characteristics among the examined clays. Ka possesses a highly crystalline structure, the smallest particle size (D_{50} = 8.5 μm), and the highest Al_2O_3 content (39.50%). Its unique dehydroxylation temperature range (700–750 °C) and substantial bound portlandite formation, as indicated by the Chapelle test, affirm its superior pozzolanic reactivity. Mn, a 2:1 clay mineral with medium crystallinity, shows good reactivity. It has a slightly smaller particle size (6.1 μm) and dehydroxylates within a similar range (700–750 °C) but contains lower Al_2O_3 (27.13%) and higher K_2O (2.55%), which may limit its reactivity compared to Ka. IL, also a 2:1 clay, has the largest particle size (16.8 μm), lower crystallinity, and a less favorable oxide composition for reactivity, particularly its lower Al_2O_3 (18.64%) and higher Fe_2O_3 (7.13%). It dehydroxylates at a higher temperature (800–850 °C), and the reduced bound portlandite observed post-calcination reflects its relatively average pozzolanic reactivity.

Table 3. Comparative analysis of the raw (R) and activated calcined (C) clays.

Clay Type	Clay Structure	PSD	XRD	TGA	Blain Fineness	Weight Loss (%)	Oxide Content (%)					MCT	Reactivity of the Clay
		D50 µm	Crystallinity	Dehydroxylation	cm^2/g	LOI	Al_2O_3 Content	SiO_2 Content	K_2O Content	CaO Content	Fe_2O_3 Content	Bound Portlandite Content	
Ka (R)	1:1	4.5	Highly crystalline	700–750 °C	6552	12.15	37.91	48.08	0.04	3.15	1.32	--	Excellent
Ka (C)	Disorder	8.5	Very low crystalline with a sharp hump at 20–40° 2θ	---	10,930	2.97	39.50	50.22	0.06	1.89	1.55	1164	
Mn (R)	2:1	4.9	Medium crystalline	700–750 °C	4545		29.23	41.45	2.55	9.60	2.32	---	Good
Mn (C)	Disorder	6.1	Low crystalline		10,000		27.13	41.45	2.55	8.76	2.02	881	
IL (R)	2:1	11	Low crystals	800–850 °C	4650	10.29	19.57	66.40	1.49	4.73	6.98	----	Average
IL (C)	Partially disorder	16.8	Resembles raw clay structure		8765	0.89	18.64	64.06	1.57	4.64	7.13	783	

The main focus of this research is the impact of these clays on the performance of LC^3. However, the characterization conducted after calcination, such as reduced crystallinity (XRD), dehydroxylation behavior (TGA), and oxide content, offers valuable insights into the inherent reactivity of each clay. These findings correspond with the results of the Chapelle test and help clarify the differences observed in hydration and strength development. These variations highlight how structural order [38], particle fineness [39], oxide composition [40], and thermal behavior together affect the reactivity potential of clays in LC^3 formulations. The mineralogical and compositional characteristics are anticipated to substantially affect the hydration mechanisms and strength development in LC^3 systems, as examined in the subsequent sections.

3.2. Heat of Hydration

Figure 9 displays the cumulative heat of hydration (kJ/kg) for OPC and the various LC^3 blends over the hydration period. Heat of hydration was measured on cement paste using an isothermal conduction calorimeter by ASTM C1702 (2017). The values are reported per kilogram of total binder (including clinker, calcined clay, limestone, and gypsum), ensuring a fair comparison between OPC and LC^3 systems despite the difference in clinker content. OPC shows the highest total heat release, mainly due to its higher clinker content, which is only partially replaced in LC^3 systems [41]. Among the various LC^3 blends, LC^3-Ka showcases the highest initial heat release during the first three days of hydration, attributed to the rapid pozzolanic reactions of metakaolin. This early activity underscores the high reactivity of calcined kaolinite [42]. However, after approximately three days, the heat release from LC^3-Ka experiences a notable slowdown. By day five, its cumulative heat release drops below that of the other LC^3 blends. Interestingly, the heat evolution of LC^3-Ka begins to rise again around the seventh day, indicating a secondary phase of reactivity. The initial spike in heat release, followed by a decline and subsequent increase for LC^3-Ka, reflects a typical two-phase hydration behavior in LC^3 systems. The early peak is driven by the swift reactivity of metakaolin and clinker hydration, while the temporary reduction in heat release suggests a decreased availability of portlandite. The increase in heat evolution after five days is linked to the delayed formation of carboaluminate phases and ongoing pozzolanic activity, which is aligned with previous studies [34,43]. LC^3-IL, despite its lower pozzolanic reactivity, demonstrates a slightly higher cumulative heat release than LC^3-Mn over the same time frame [44]. This discrepancy may be attributed to variations in particle morphology or packing density, which can influence the accessibility of water and clinker particles during the hydration process. These findings emphasize the significance of clay mineralogy in shaping hydration kinetics, as both early and later-stage reactions differ based on the type of calcined clay utilized in the LC^3 system.

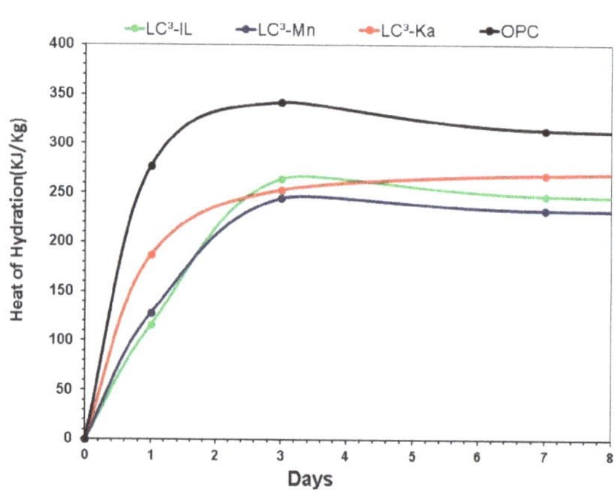

Figure 9. Heat of hydration of OPC and LC^3 clay blends.

3.3. Hydration Products

3.3.1. X-Ray Diffraction (XRD)

The hydration products were examined using XRD techniques. The type of calcined clay in LC^3 significantly impacts the hydration products formed. Figure 10 presents the XRD spectra for the OPC and LC^3 systems after 7 and 28 days of hydration. The hydration products formed in LC^3 blends depend on the types of calcined clay present [45].

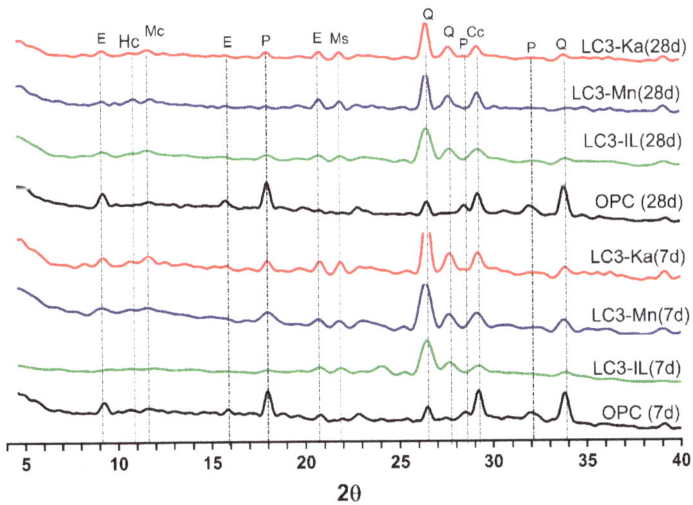

Figure 10. XRD assessment of LC^3 and OPC clay blends at the age of 7 and 28 days.

After seven days, the primary phases detected in OPC via XRD are portlandite and ettringite [46]. The peaks become more intense with increasing age. In contrast, the ternary binders LC^3-Ka, LC^3-Mn, and LC^3-IL also show the presence of hemicarboaluminate (Hc) and monocarboaluminate (Mc), due to the interaction of calcined clay and limestone, which promotes carbonate formation. These peaks become more pronounced with age, as seen in the 28-day results [47]. Although all LC^3 systems generated comparable hydration products, their kinetics and intensities exhibited significant variation depending on the mineralogy of the calcined clay. The hemi and monocarboaluminate peaks are prominent in LC^3-Ka, followed by LC^3-Mn and LC^3-IL. LC^3-IL exhibits the lowest peak intensity due to its slower pozzolanic reaction, as indicated by its lower compressive strength and reactivity results. However, the OPC spectra exhibit higher-intensity ettringite and portlandite peaks than the LC^3 blends [48]. The intensity of the portlandite peaks decreases with age thanks to the limestone and calcined clay pozzolanic reaction. This is also due to LC^3's secondary hydration reactions and clinker dilution effect [45,49]. However, the decrease is slower in LC^3-IL, though all binders eventually reach similar levels. The limited development of the Mc and Hc phases in LC3-IL corresponds with its lower 28-day compressive strength, as a deficiency in reactive alumina restricts the potential for advantageous secondary reactions. This relationship underscores that the mineralogical properties of clay minerals directly influence both phase development and mechanical performance. Distinct peaks of quartz (Q) are observed at approximately 26.6° 2θ, and those of calcite (Cc) at around 29.4° 2θ are more prominent in LC^3 systems due to the presence of unreacted phases from calcined clay and limestone filler. These peaks serve as indirect evidence of the degree of reaction. Notably, LC3-IL exhibits more pronounced Q peaks at 28 days, indicating lower pozzolanic reactivity, which aligns with the mechanical performance data [49].

The hydration products from different ternary binders were similar, but the hydration behaviors varied based on the mineralogy of calcined clay employed. These findings emphasize that not all calcined clays exhibit uniform reactivity, and that mineralogy is key in optimizing LC^3 performance, particularly in formulations where early strength and durability are essential. The pozzolanic and secondary hydration reactions in non-kaolinitic LC^3 blends are determined to exhibit comparable efficiency to those involving kaolinite.

3.3.2. Thermogravimetric Analysis (TGA)

The TGA results were differentiated to obtain the derivative thermogravimeter (DTG) curve, which is shown in Figure 11. The curve mainly has three weight-loss regions. The weight loss between 150 °C and 300 °C is attributed to the decomposition of ettringite, carbonates (Mc and Hc), and C-(A)-S-H [50]. The decomposition of portlandite ($Ca(OH)_2$ (CH) results in a peak between 400 °C and 500 °C [51]. Weight loss above 600 °C results from the disintegration of calcium carbonate ($CaCO_3$) [52].

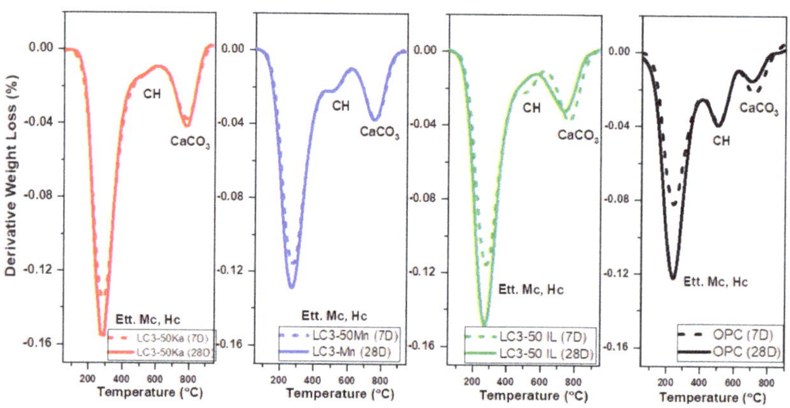

Figure 11. DTG of OPC and LC^3 clay blends at 7 and 28 days.

As hydration progresses, ettringite, Mc, Hc, and C-S-(A)-H increase significantly in all the blends. However, due to the presence of Hc and Mc phases, the loss in LC^3 blends is significant. In the OPC sample, there is also a substantial increase in CH. However, in the LC^3 mixtures, except for LC^3-IL, the weight loss due to CH significantly decreases after 28 days. This decrease is due to the pozzolanic reaction between calcined clay and portlandite, which forms additional C-A-S-H gel and consumes CH in this process. This trend is specifically observable in LC^3-Ka and LC^3-Mn blends, where the reactivity of kaolinite and montmorillonite causes a significant decrease in the CH peak. Conversely, the lower reactivity of illite in LC^3-IL results in a more gradual pozzolanic reaction, elucidating the relatively minor fluctuation in CH content over time. These variations underscore the mineralogical differences among the clays used and their respective reactivities, with kaolinite showing the highest reactivity and illite the lowest. This interpretation aligns with the XRD and compressive strength results, reinforcing the notion that kaolinite- and montmorillonite-rich clays facilitate more rapid portlandite consumption, while illite-rich clays exhibit a slower reaction. Moreover, the diminished availability of portlandite for carbonation results in a gradual decline in the quantity of $CaCO_3$ produced in LC^3 blends, unlike the OPC system, where CH persists and thus enhances $CaCO_3$ generation. In LC^3, the blend initially has a higher calcite content due to the limestone in the binder [53]. However, as hydration progressed, the observed calcite content reduced because there was less portlandite available for carbonation as a result of its gradual consumption in the pozzolanic reaction [14].

3.3.3. Scanning Electron Microscopy (SEM)

The SEM micrographs in Figure 12 illustrate the morphological differences between the hydration products of OPC and those generated in LC^3 systems with various clay types at the age of 28 days. The OPC matrix exhibits a mostly crystalline microstructure, distinguished by many plate-like CH crystals and needle-shaped ettringite/AFm phases, interspersed within a matrix of C-S-H and secondary aluminous hydrates [54]. This illustrates the traditional hydration process characterized by the development of portlandite and ettringite [55].

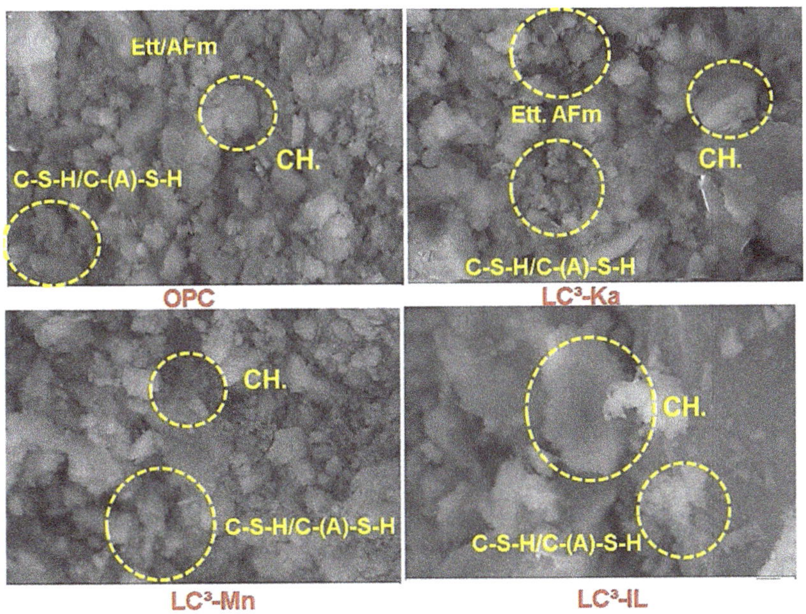

Figure 12. SEM analysis of OPC and all LC^3-blends (LC^3-Ka, LC^3-Mn, LC^3-IL).

The LC^3-Ka system demonstrates a denser and more homogeneous microstructure, characterized by a notable reduction in visible portlandite and a fine distribution of C-S-H/C-(A)-S-H phases [56]. The reduced presence of CH indicates the increased pozzolanic activity of kaolinite, which actively consumes portlandite during secondary hydration reactions. The microstructural refinement in LC^3-Ka correlates with its enhanced mechanical performance and reactivity, as demonstrated by TGA and compressive strength findings.

The LC^3-Mn sample exhibits a compact structure characterized by well-developed C-S-H/C-(A)-S-H gels and moderate CH content [57]. Montmorillonite, despite having a lower alumina content than kaolinite, likely continues to dissolve over time due to its layered 2:1 structure, which aids in the formation of hydration products that are morphologically similar to those found in kaolinite-based blends [57]. The mechanical data indicate that LC^3-Mn attained strength values comparable to those of LC^3-Ka at 28 days.

The LC^3-IL blend exhibits a more porous and less cohesive microstructure characterized by discrete CH crystals and a limited distribution of C-S-H gels. The existence of unreacted or partially reacted phases indicates that illite, although a low-cost and readily available material, demonstrates slower reaction kinetics and reduced pozzolanic reactivity at early ages [58]. The gradual development of hydration products over time, as evidenced by the progression of compressive strength, suggests that illite is a viable supplementary cementitious material for long-term or non-structural applications. Nevertheless, the incremental formation of hydration products. However, the progressive formation of hydration products over time, as evidenced by the increase in compressive strength, suggests that illite continues to be an effective supplementary cementitious material in long-term or non-structural applications.

The SEM analysis supports the TGA and XRD analysis and chemical and mechanical properties of the subsequent clays, demonstrating that the pozzolanic potential of the clays is evident in both bulk hydration behavior and the resulting microstructural development. The reduction in portlandite and the densification of the binder matrix in LC^3 systems, especially with kaolinite and montmorillonite, indicate improved durability and mechanical integrity in blended cement formulations.

3.4. Setting Time and Consistency of Blends

Figure 13 displays the findings from the investigations on normal consistency and setting time. All LC^3 binders demonstrated slightly superior consistency compared to OPC, with LC^3-IL achieving the highest value. This improvement in consistency can be attributed to the finer particle size and increased surface area of the calcined clay particles, resulting in greater water demand [59]. Among the various blends, LC^3-IL required a larger volume of water to achieve standard consistency, likely due to the platy morphology and relatively low pozzolanic activity of illite, which impedes early hydration reactions.

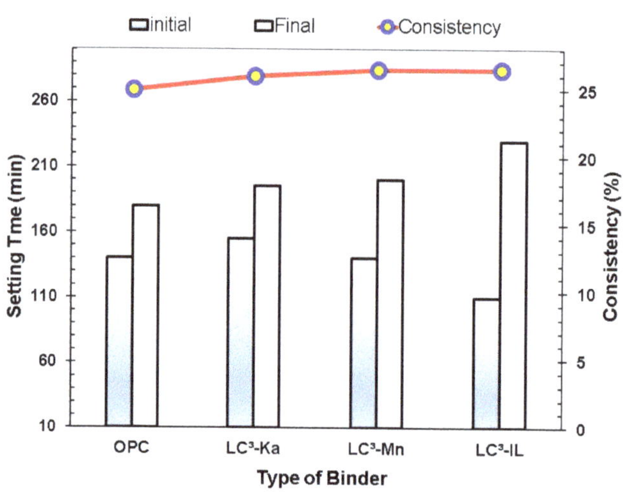

Figure 13. Consistency and setting time of the OPC and LC^3 clay blends.

In terms of setting time, all LC^3 binders exhibited longer initial and final setting times compared to OPC. This trend is consistent with previous findings that the incorporation of calcined clay and limestone tends to dilute the clinker content and retard the formation of C-S-H gel in the early age [57,60]. Among the various blends, LC^3-IL required a larger volume of water to achieve standard consistency, likely due to the platy morphology and relatively low pozzolanic activity of illite, which impedes early hydration reactions. In contrast, LC^3-Ka and LC^3-Mn showed relatively moderate increases in setting time, owing to the higher reactivity of kaolinite and montmorillonite, which promote earlier nucleation and growth of hydrates. The mineralogical composition, degree of reactivity, and physical properties of the calcined clays utilized in LC^3 systems significantly affect the setting time and consistency [61]. However, despite these variations, all LC^3 mixtures meet the minimum setting time criteria outlined in ASTM C191.

3.5. Compressive Strength

The development of compressive strength in LC^3 mortars at both 7 and 28 days, as illustrated in Figure 14, reveals notable performance trends influenced by the type of calcined clay utilized. To aid comparison, the 7-day and 28-day compressive strength values of the OPC mix are also indicated as horizontal reference lines in Figure 13. Among the ternary blends, LC^3-Ka exhibited the highest compressive strength, reaching 20.4 MPa at 7 days and 30.8 MPa at 28 days, closely competing with OPC, which recorded values of 23.0 MPa and 31.9 MPa for the respective timeframes. This strong early-age performance can be attributed to the high reactivity of kaolinite, a 1:1 layered silicate mineral, which accelerates the pozzolanic reaction and promotes the rapid consumption of portlandite [62]. However, it was slightly lower than OPC due to its lower clinker content and dilution effect. The acceleration caused by the clay was insufficient to offset the dilution effect [63]. LC^3-Mn, composed of montmorillonite clay, exhibited compressive strengths of 17.9 MPa at 7 days and 27.9 MPa at 28 days, slightly inferior to LC^3-Ka but demonstrating significant temporal enhancement. The 2:1 layered structure of montmorillonite promotes gradual secondary hydration [30], consistent with the delayed yet consistent formation of hydration products indicated by XRD and DTG analyses (see Figures 10 and 11). The moderate decrease in CH content from 7 to 28 days, as evidenced by TGA results, corroborates this postponed pozzolanic effect.

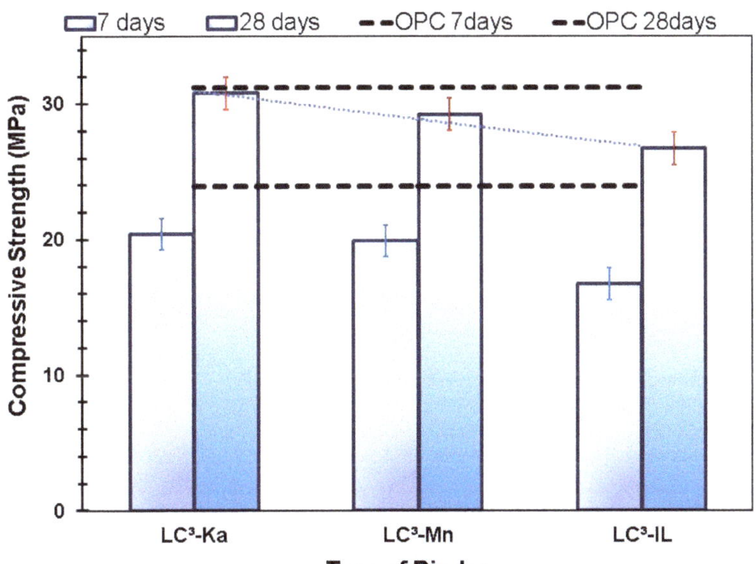

Figure 14. Compressive strength of LC^3 mortars containing kaolinite (LC^3-Ka), montmorillonite (LC^3-Mn), and illite (LC^3-IL) at 7 and 28 days. The black horizontal lines indicate the corresponding compressive strength values of the OPC control (at 7 and 28 days).

In contrast, LC^3-IL, which includes illite clay, demonstrated the lowest strength at an early age, measuring 16.3 MPa at 7 days, but experienced a significant increase to 23.9 MPa by 28 days. This approximately 70% enhancement suggests that, despite illite's relatively inert nature due to its partially crystalline and impure structure, it still contributes to long-term strength development through gradual pozzolanic reactions. The XRD spectra confirmed a reduction in carboaluminate peaks in LC^3-IL, while the TGA results showed minimal consumption of calcium hydroxide at an early age, both of which support the notion of delayed pozzolanic activity. Nonetheless, the illite-based LC^3 achieved approximately 75% of the strength of OPC by 28 days, indicating that even low-reactivity clays can significantly enhance long-term performance [63]. The observed trends underscore the significant influence of clay mineralogy on hydration kinetics and strength development in LC^3 systems. The high reactivity of kaolinite contributes to enhanced early strength, while montmorillonite offers moderate but consistent strength gains. In contrast, illite primarily aids strength development at later ages. The interaction between limestone and various clay types, evidenced by the formation of carboaluminates and the consumption of calcium hydroxide, underscores the need for customized LC^3 formulations to optimize performance. Pearson's correlation coefficient (R = 0.998) was determined between the portlandite fixation values (modified Chapelle test) and the 28-day compressive strengths of LC^3 mortars presented in Figure 15, indicating a nearly linear positive association between pozzolanic reactivity and mechanical performance.

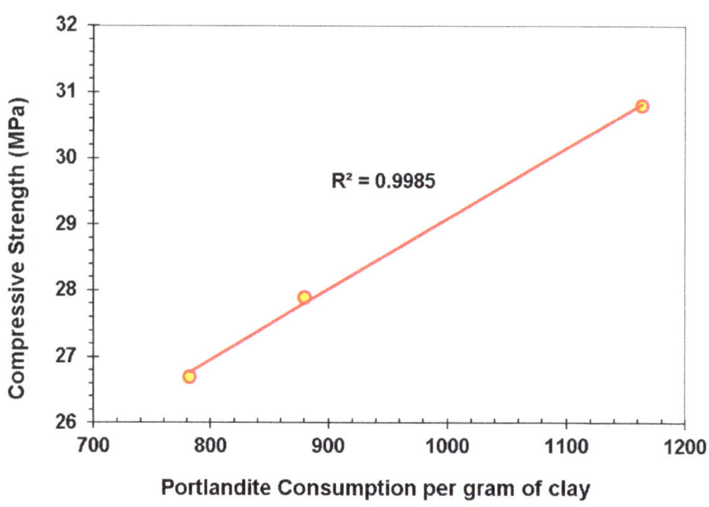

Figure 15. Correlation between 28-day compressive strength of LC^3 blends and pozzolanic reactivity of calcined clays as determined by the modified Chapelle test.

3.6. Overall Performance Summary

Figure 16 displays a radar chart that compares the normalized performance of different LC^3 blends—specifically, LC^3-Ka, LC^3-Mn, and LC^3-IL—against OPC across seven key parameters. All parameters have been normalized based on the highest value recorded among all mixes, allowing for direct visual comparison. This approach emphasizes performance distribution instead of absolute values, complementing the detailed trends reported earlier. LC^3-Ka demonstrated consistently high performance, closely aligning with OPC in compressive strength, setting time, and pozzolanic reactivity, underscoring kaolinite's suitability for LC^3 production. LC^3-Mn exhibited moderate results across most indicators, while LC^3-IL, despite lower early strength and reactivity, performed well in terms of water demand and setting characteristics. These performance trends reflect the underlying mineralogical differences among the clays. While all three clay types are viable for LC^3 applications, kaolinite offers the most balanced and structurally robust performance. Illite, though less reactive, remains a promising low-cost alternative for non-structural or blended cement uses.

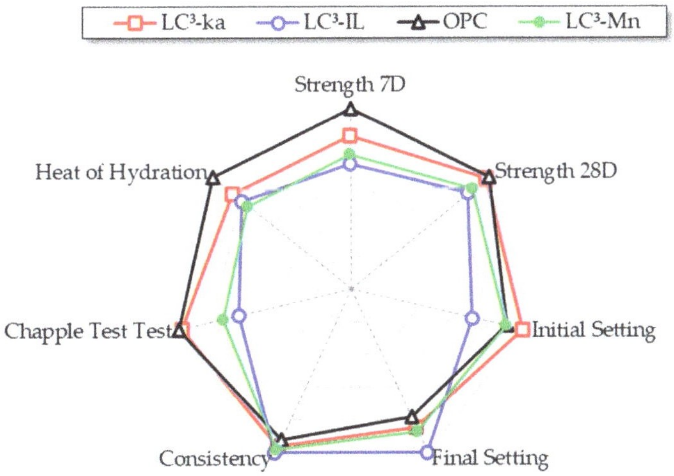

Figure 16. Radar graph showing the overall performance.

4. Conclusions

This study examined the classification and applicability of diverse low-grade clays in LC^3 systems, emphasizing their thermal properties, reactivity, and contribution to mechanical performance. The results indicated that optimal calcination temperatures are specific to each clay type, with kaolinite and montmorillonite undergoing effective transformation at 750 °C, whereas impure illite necessitated higher thermal energy for complete dehydroxylation.

The findings from XRD, TGA, and reactivity assessments revealed a significant correlation between the degree of dehydroxylation and the mechanical behavior of the LC^3 blends. Kaolinite-based LC^3 consistently demonstrated superior performance during both early and later curing stages, highlighting its potential as a dependable supplementary SCM, even in its low-grade form. Montmorillonite also exhibited encouraging results, whereas illite, despite a slight deficiency in early-age strength, showed satisfactory strength development over time.

The modified Chapelle test has proven to be a reliable indicator of clay reactivity, providing practical support for the early screening of clay sources. This study emphasizes that, rather than focusing solely on low-grade clays, widely available low-cost non-premium clays can be significant components in sustainable cement formulations when processed accurately. Optimizing blend composition and tailoring calcination parameters are essential for utilizing these resources in low-carbon construction materials.

Despite demonstrating the highest strength and reactivity, kaolinite is frequently costly and unavailable in many regions of the Southern Hemisphere. Montmorillonite and illite, due to their cost-effectiveness and local abundance, with illite being the least expensive, present viable options. Montmorillonite, specifically, attained equal strength at 28 days and can serve as an efficient substitute for kaolinite when early strength is not critical.

Author Contributions: Conceptualization, N.B.; methodology, S.N.H.K. and N.B.; validation, A.K.; formal analysis, N.A.; data curation, K.J.; writing—original draft preparation, S.N.H.K., I.A.K. and M.J.S.; writing—review and editing, I.A.K., K.J., S.A. and A.K. All authors have read and agreed to the published version of the manuscript.

Funding: This research received no external funding.

Data Availability Statement: Data will be provided by the corresponding authors on request.

Conflicts of Interest: The authors declare no conflicts of interest.

Abbreviations

The following abbreviations are used in this manuscript:

LC^3	Limestone calcined clay cement (LC^3)
OPC	Ordinary Portland Cement (OPC)
IL	Illite

References

1. Py, L.G.; Neto, J.S.A.; Longhi, M.A.; Kirchheim, A.P. Evaluation of ultrafine calcined clays on LC^3 cements on the sulfate requirement, water demand and strength development. *Mater. Struct.* **2024**, *57*, 13. [CrossRef]
2. Scrivener, K.; Avet, F.; Maraghechi, H.; Zunino, F.; Ston, J.; Hanpongpun, W.; Favier, A. Impacting factors and properties of limestone calcined clay cements (LC^3). *Green Mater.* **2019**, *7*, 3–14. [CrossRef]
3. Sun, J.; Zunino, F.; Scrivener, K. Hydration and phase assemblage of limestone calcined clay cements (LC^3) with clinker content below 50%. *Cem. Concr. Res.* **2024**, *177*, 107417. [CrossRef]
4. Kafodya, I.; Basuroy, D.; Marangu, J.M.; Kululanga, G.; Maddalena, R.; Novelli, V.I. Mechanical performance and physico-chemical properties of limestone calcined clay cement (LC^3) in Malawi. *Buildings* **2023**, *13*, 740. [CrossRef]
5. Sui, H.; Hou, P.; Liu, Y.; Sagoe-Crentsil, K.; de Souza, F.B.; Duan, W. Limestone calcined clay cement: Mechanical properties, crystallography, and microstructure development. *J. Sustain. Cem. Mater.* **2023**, *12*, 427–440. [CrossRef]
6. Yu, J.; Mishra, D.K.; Hu, C.; Leung, C.K.; Shah, S.P. Mechanical, environmental and economic performance of sustainable Grade 45 concrete with ultrahigh-volume Limestone-Calcined Clay (LCC). *Resour. Conserv. Recycl.* **2021**, *175*, 105846. [CrossRef]
7. Brigatti, M.F.; Galán, E.; Theng, B.K.G. structure and mineralogy of clay minerals. In *Developments in Clay Science*, 2nd ed.; Bergaya, F., Lagaly, G., Eds.; Elsevier: Amsterdam, The Netherlands, 2013; Volume 5, pp. 21–81. [CrossRef]
8. Fernandez, R.; Martirena, F.; Scrivener, K.L. The origin of the pozzolanic activity of calcined clay minerals: A com-parison between kaolinite, illite and montmorillonite. *Cem. Concr. Res.* **2011**, *41*, 113–122. [CrossRef]
9. Li, K.J.; Kurniawan, A.; Christidis, G.E.; He, J.Y.; Zhou, C.H. Interfacial interactions controlling adsorption of metal cations on montmorillonite. *Am. Miner.* **2024**, *109*, 633–655. [CrossRef]
10. Kumari, N.; Mohan, C. Basics of clay minerals and their characteristic properties. *Clay Clay Min.* **2021**, *24*, 1–29.
11. Ghosh, B.; Chakraborty, D. An overview of the clay minerals: Composition, classification, internal structure and properties. In *Clay Minerals: Their Antimicrobial and Antitoxic Applications*; Springer Nature: Berlin/Heidelberg, Germany, 2023; pp. 1–24.
12. Hay, R.; Celik, K. Performance enhancement and characterization of limestone calcined clay cement (LC3) produced with low-reactivity kaolinitic clay. *Constr. Build. Mater.* **2023**, *392*, 131831. [CrossRef]
13. Hay, R.; Celik, K. Comparative performance of OPC and LC^3-based composites under combined action of carbonation and chloride exposure. *Constr. Build. Mater.* **2024**, *429*, 136428. [CrossRef]
14. Her, S.; Im, S.; Liu, J.; Suh, H.; Kim, G.; Sim, S.; Wi, K.; Park, D.; Bae, S. Exploring the potential of pulverized oyster shell as a limestone substitute in limestone calcined clay cement (LC^3) and its implications for performance. *Constr. Build. Mater.* **2024**, *425*, 135918. [CrossRef]
15. Han, F.; Li, Y.; Zhu, Z.; Zhang, Z. Synergistic utilization of Al-rich and carbonate-rich mineral admixtures toward sustainable low-carbon cementitious materials: A review. *J. Build. Eng.* **2025**, *105*, 112523. [CrossRef]
16. Diaz, A.A.; Reyes, R.S.A.; Hanein, T.; Irassar, E.F.; Juenger, M.; Kanavaris, F.; Maier, M.; Marsh, A.T.; Sui, T.; Thienel, K.-C.; et al. Properties and occurrence of clay resources for use as supplementary cementitious materials: A paper of RILEM TC 282-CCL. *Mater. Struct.* **2022**, *55*, 139. [CrossRef]
17. Blouch, N.; Rashid, K.; Zafar, I.; Ltifi, M.; Ju, M. Prioritization of low-grade kaolinite and mixed clays for performance evaluation of Limestone Calcined Clay Cement (LC^3): Multi-criteria assessment. *Appl. Clay Sci.* **2023**, *243*, 107080. [CrossRef]

18. Krishnan, S.; Rao, D.G.; Bishnoi, S. Why Low-Grade Calcined Clays Are the Ideal for the Production of Limestone Calcined Clay Cement (LC^3). In *Calcined Clays for Sustainable Concrete: Proceedings of the 3rd International Conference on Calcined Clays for Sustainable Concrete, New Delhi, India, 15–17 October 2019*; Springer Nature: Berlin/Heidelberg, Germany, 2020; Volume 25, p. 125.
19. Zolfagharnasab, A.; Ramezanianpour, A.A.; Bahman-Zadeh, F. Investigating the potential of low-grade calcined clays to produce durable LC^3 binders against chloride ions attack. *Constr. Build. Mater.* **2021**, *303*, 124541. [CrossRef]
20. De Maio, U.; Gaetano, D.; Greco, F.; Lonetti, P.; Pranno, A. An adaptive cohesive interface model for fracture propagation analysis in heterogeneous media. *Eng. Fract. Mech.* **2025**, *325*, 111330. [CrossRef]
21. Tijssens, M.; Sluys, L.; van der Giessen, E. Simulation of fracture of cementitious composites with explicit modeling of microstructural features. *Eng. Fract. Mech.* **2001**, *68*, 1245–1263. [CrossRef]
22. Werling, N.; Kaltenbach, J.; Weidler, P.G.; Schuhmann, R.; Dehn, F.; Emmerich, K. Solubility of Calcined Kaolinite, Montmorillonite, and Illite in High Molar NaOH and Suitability as Precursors for Geopolymers. *Clays Clay Miner.* **2022**, *70*, 270–289. [CrossRef]
23. Alujas, A.; Fernández, R.; Quintana, R.; Scrivener, K.L.; Martirena, F. Pozzolanic reactivity of low grade kaolinitic clays: Influence of calcination temperature and impact of calcination products on OPC hydration. *Appl. Clay Sci.* **2015**, *108*, 94–101. [CrossRef]
24. Kassa, A.E.; Shibeshi, N.T.; Tizazu, B.Z. Kinetic analysis of dehydroxylation of Ethiopian kaolinite during calcination. *J. Therm. Anal. Calorim.* **2022**, *147*, 12837–12853. [CrossRef]
25. Irassar, E.F.; Bonavetti, V.L.; Castellano, C.C.; Trezza, M.A.; Rahhal, V.F.; Cordoba, G.; Lemma, R. Calcined illite-chlorite shale as supplementary cementing material: Thermal treatment, grinding, color and pozzolanic activity. *Appl. Clay Sci.* **2019**, *179*, 105143. [CrossRef]
26. Ferreiro, S.; Canut, M.; Lund, J.; Herfort, D. Influence of fineness of raw clay and calcination temperature on the performance of calcined clay-limestone blended cements. *Appl. Clay Sci.* **2019**, *169*, 81–90. [CrossRef]
27. Waheed, I.F.; Awsaj, M.M.; Dahham, O.S.; Jabbar, M.Q.; Al-Abady, F.M.; Al-Samarrai, M.A.F. Synthesis of CoMnFe2O4 hollow microstructure decorated GO for photocatalytic degradation of organic dyes. *Arab. J. Chem.* **2024**, *17*, 105934. [CrossRef]
28. Zhao, Y.; Qiu, J.; Guo, Z.; Zhang, S.; Wu, P.; Sun, X. Activation the hydration properties of illite-containing tailings to prepare a binder for cemented paste backfill. *Constr. Build. Mater.* **2022**, *318*, 125989. [CrossRef]
29. Marsh, A.T.; Krishnan, S.; Bernal, S.A. Structural features of thermally or mechano-chemically treated montmorillonite clays as precursors for alkali-activated cements production. *Cem. Concr. Res.* **2024**, *181*, 107546. [CrossRef]
30. Rakhimova, N. Montmorillonite clays in Portland clinker-reduced, non-clinker cements, and cement composites: A review. *Constr. Build. Mater.* **2024**, *411*, 134678. [CrossRef]
31. Wang, S.; Gainey, L.; Mackinnon, I.D.; Allen, C.; Gu, Y.; Xi, Y. Thermal behaviors of clay minerals as key components and additives for fired brick properties: A review. *J. Build. Eng.* **2023**, *66*, 105802. [CrossRef]
32. Zunino, F.; Scrivener, K. Reactivity of kaolinitic clays calcined in the 650 °C–1050 °C temperature range: Towards a robust assessment of overcalcination. *Cem. Concr. Compos.* **2024**, *146*, 105380. [CrossRef]
33. Luzu, B.; Duc, M.; Djerbi, A.; Gautron, L. High performance illitic clay-based geopolymer: Influence of thermal/mechanical activation on strength development. *Appl. Clay Sci.* **2024**, *258*, 107445. [CrossRef]
34. Scrivener, K.; Martirena, F.; Bishnoi, S.; Maity, S. Calcined clay limestone cements (LC^3). *Cem. Concr. Res.* **2018**, *114*, 49–56. [CrossRef]
35. *ASTM C 191-92*; Standard Test Method for Time of Setting of Hydraulic Cement by Vicat Needle, Annual Book of ASTM. ASTM International: West Conshohocken, PA, USA, 1995; Volume 4.
36. *ASTM C 191-08*; Standard Test Methods for Time of Setting of Hydraulic Cement by Vicat Needle. ASTM International: West Conshohocken, PA, USA, 2008.
37. Chen, C.; Zhang, Y.; Li, X.; He, J.; Gao, F.; Chen, Z. Investigations into methane hydrate formation, accumulation, and distribution in sediments with different contents of illite clay. *Appl. Energy* **2024**, *359*, 122661. [CrossRef]
38. Yaman, İ.Ö. *Effects of Clay Type on the Properties of Limestone Calcined Clay Cement*; Middle East Technical University: Ankara, Turkey, 2024.
39. Allegretta, I.; Pinto, D.; Eramo, G. Effects of grain size on the reactivity of limestone temper in a kaolinitic clay. *Appl. Clay Sci.* **2016**, *126*, 223–234. [CrossRef]
40. Fatima, M.; Rashid, K.; Blouch, N. Thermo-Mechano-Chemo Dynamics of Clay's Reactivity and its Correlation with Limestone Calcined Clay Cement (LC^3) Performance. *J. Adv. Concr. Technol.* **2024**, *22*, 689–705. [CrossRef]
41. Barbhuiya, S.; Nepal, J.; Das, B.B. Properties, compatibility, environmental benefits and future directions of limestone calcined clay cement (LC^3) concrete: A review. *J. Build. Eng.* **2023**, *79*, 107794. [CrossRef]
42. Chen, Z.; Luo, K.; Ye, H. Understanding reactivity of calcined marine clay as a supplementary cementitious material through structural transformation of clay minerals. *Cem. Concr. Compos.* **2025**, *160*, 106066. [CrossRef]

43. Antoni, M.; Rossen, J.; Martirena, F.; Scrivener, K. Cement substitution by a combination of metakaolin and limestone. *Cem. Concr. Res.* **2012**, *42*, 1579–1589. [CrossRef]
44. Blouch, N.; Rashid, K.; Ju, M. Exploring low-grade clay minerals diving into limestone calcined clay cement (LC^3): Characterization—Hydration—Performance. *J. Clean. Prod.* **2023**, *426*, 139065. [CrossRef]
45. Boakye, K.; Khorami, M. Performance of Calcined Impure Kaolinitic Clay as a Partial Substitute for Portland Cement Concrete: A Review. *J. Compos. Sci.* **2025**, *9*, 145. [CrossRef]
46. Adem, J.K.; Seo, J.; Park, S.; Kim, G. Effects of low-lime calcium silicate cement addition on the hydration, carbonation, and microstructural characteristics of OPC pastes by carbonation curing. *J. Build. Eng.* **2025**, *107*, 112683. [CrossRef]
47. Jayathilakage, R.; Gunasekara, C.; Law, D.; Setunge, S. A combination technique to improve natural low-grade illite as supplementary cementitious material for concrete. *Constr. Build. Mater.* **2025**, *477*, 141334. [CrossRef]
48. Bansal, T.; Talakokula, V. Comparative analysis of very-early age hydration process: LC^3 vs. conventional and blended cement pastes using embedded piezo sensors. *Measurement* **2024**, *229*, 114433. [CrossRef]
49. Shao, Z.; Cao, M. Hydration mechanism of limestone calcined clay cement containing calcined coal gangue. *Constr. Build. Mater.* **2024**, *438*, 136906. [CrossRef]
50. Shen, X.; He, H.; He, C.; Li, B.; Luo, W.; Ren, P. Low-carbon blended cement containing wet carbonated municipal solid waste incineration fly ash and mechanically activated coal fly ash. *Case Stud. Constr. Mater.* **2024**, *21*, e03671. [CrossRef]
51. Ponomar, V.; Yliniemi, J.; Kilpimaa, K. Influence of calcium oxide and hydroxide on non-ferrous metallurgical slag hydration in alkaline environments. *Cem. Concr. Compos.* **2024**, *152*, 105673. [CrossRef]
52. Ren, S.; Huang, G.; Zhang, W.; Sha, X.; Liu, G.; Lin, R.-S.; Chen, L. Enhancing cement-based materials hydration and carbonation efficiency with pre-carbonated lime mud. *J. CO2 Util.* **2024**, *88*, 102928. [CrossRef]
53. Mañosa, J.; Calderón, A.; Salgado-Pizarro, R.; Maldonado-Alameda, A.; Chimenos, J.M. Research evolution of limestone calcined clay cement (LC^3), a promising low-carbon binder—A comprehensive overview. *Heliyon* **2024**, *10*, e25117. [CrossRef]
54. Wu, H.; He, M.; Cheng, J.; Wang, T.; Che, Y.; Du, Y. Study on the synergistic effects of OPC and silica fume on the mechanical and microstructural properties of geopolymer mortar. *Constr. Build. Mater.* **2024**, *434*, 136740. [CrossRef]
55. Georget, F.; Wilson, W.; Scrivener, K.L. Edxia: Microstructure characterisation from quantified SEM-EDS hypermaps. *Cem. Concr. Res.* **2021**, *141*, 106327. [CrossRef]
56. Atasever, M.; Erdoğan, S.T. Effects of clay type and component fineness on the hydration and properties of limestone calcined clay cement. *Mater. Struct.* **2024**, *57*, 183. [CrossRef]
57. Hu, Y.; Xiong, L.; Yan, Y.; Geng, G. Performance of limestone calcined clay cement (LC^3) incorporating low-grade marine clay. *Case Stud. Constr. Mater.* **2024**, *20*, e03283. [CrossRef]
58. Rehman, M.U.; MacLeod, A.J.; Gates, W.P. Phase development and mechanical strength of limestone calcined clay cement utilising Australian bentonite and plasterboard waste. *Constr. Build. Mater.* **2024**, *445*, 137937. [CrossRef]
59. Ferreiro, S.; Herfort, D.; Damtoft, J. Effect of raw clay type, fineness, water-to-cement ratio and fly ash addition on workability and strength performance of calcined clay—Limestone Portland cements. *Cem. Concr. Res.* **2017**, *101*, 1–12. [CrossRef]
60. Zhang, J.; Luo, Q.; Zhang, X. Exploring the influence of calcined clay grade on the rheological dynamics of LC^3 mortar. *Constr. Build. Mater.* **2024**, *444*, 137852. [CrossRef]
61. Zhao, Y.; Zhang, Y. A Review on Hydration Process and Setting Time of Limestone Calcined Clay Cement (LC^3). *Solids* **2023**, *4*, 24–38. [CrossRef]
62. Gu, H.; Meng, Z.; Wang, Y.; Gao, X.; Wang, R.; Wang, D.; Sheng, J.; Wang, J. Investigations into Replacing Calcined Clay with Sewage Sludge Ash in Limestone Calcined Clay Cement (LC3). *Materials* **2025**, *18*, 782. [CrossRef]
63. da Silva, M.R.C.; Andrade Neto, J.d.S.; Walkley, B.; Kirchheim, A.P. Effects of kaolinite and montmorillonite calcined clays on the sulfate balance, early hydration, and artificial pore solution of limestone calcined clay cements (LC^3). *Mater. Struct.* **2024**, *57*, 187. [CrossRef]

materials

MDPI

Article

Reducing Cement Clinker Sintering Temperature Using Fluorine-Containing Semiconductor Waste

Bilguun Mend [1,2], Youngjun Lee [1], Jang-Ho Jay Kim [2,*] and Yong-Sik Chu [1,*]

1 Climate and Energy R&D Group, Korea Institute of Ceramic Engineering and Technology, Jinju 52581, Republic of Korea; bilguun@kicet.re.kr (B.M.); uyi047@kicet.re.kr (Y.L.)
2 School of Civil and Environmental Engineering, Yonsei University, Yonsei-ro, Seodaemun-gu, Seoul 03722, Republic of Korea
* Correspondence: jjhkim@yonsei.ac.kr (J.-H.J.K.); yschu@kicet.re.kr (Y.-S.C.)

Abstract

This study investigated the potential use of fluorine-containing semiconductor industrial sludge as a mineralizer in Portland cement clinker production. Raw mixes were prepared by partially replacing raw materials with 6%, 9%, and 12% sludge and sintered between 1300 and 1500 °C. The clinker burnability, phase composition, and chemical integrity were evaluated through FreeCaO measurements, X-ray diffraction (XRD) with Rietveld refinement, and X-ray fluorescence (XRF) analyses. The results showed that sludge addition reduced the sintering temperature by up to 150 °C, enabling near-complete clinker formation at 1300 °C for blends containing 9% and 12% sludge (FreeCaO $\leq$ 0.6 wt.% compared to 62 wt.% in the reference sample). Fluorine incorporation stabilized the reactive β–C_2S polymorph and shifted the alite (C_3S) phase distribution from stable M1 to metastable M3 and T3 phases. Additionally, the C_3A phase content decreased, and a unique fluorine-containing phase, $Al_7Ca_6O_{16}F$, formed, promoting clinker formation. Lowering the sintering temperature led to energy savings of 15–22.5% and reduced CO_2 emissions by 0.10–0.20 tons per ton of clinker, positively impacting the environment. This study demonstrates that recycling industrial sludge can enhance cement production efficiency and support environmental sustainability.

Keywords: fluorine mineralizer; semiconductor sludge; cement clinker; energy savings; CO_2 emissions

1. Introduction

The cement industry, a significant contributor to global CO_2 emissions, is in urgent need of alternative raw materials and innovative technologies. Accounting for approximately 8% of total anthropogenic releases, the industry's high temperatures (1450 °C) for clinker production demand enormous amounts of thermal energy [1,2]. The reliance on natural raw materials such as limestone and clay further exacerbates environmental impacts [3], including landscape disruption, biodiversity loss, dust emissions, and groundwater depletion. Increasing energy consumption and the depletion of limited resources underscore this urgent need for exploration. The time to act is now to reduce both the carbon footprint and the energy intensity of cement manufacturing, highlighting the importance of alternative raw materials and sustainable practices [4].

Over the past decades, various industrial by-products have been successfully utilized as partial substitutes for cement production. Fly ash, blast furnace slag, silicomanganese slag, and sewage sludge, for instance, have been incorporated into raw mixes, achieving resource conservation and reducing CO_2 emissions [5,6]. These practices align with the principles of the circular economy, demonstrating that waste streams can be transformed into valuable resources and integrated into sustainable construction practices [7,8].

Recently, semiconductor industry waste, particularly chemical mechanical polishing (CMP) sludge, has attracted increasing attention. This sludge, despite being classified as hazardous waste, holds significant potential. Its high concentrations of CaO, Al_2O_3, and fluorine (F) can significantly influence clinker mineralogy and sintering behavior [9]. The high fluorine content imparts fluxing properties that may lower the melting temperature of clinker raw mixes and promote phase formation [10,11]. This potential of semiconductor-derived fluorine-rich sludge as a mineralizing agent is a beacon of hope in the quest for sustainable cement production [12,13].

Most previous studies have focused on conventional mineralizers (such as CaF_2 or cryolite) and general industrial by-products [14,15], with limited emphasis on the mechanistic role of fluorine-rich sludge in clinker phase transformations, mineralogical stability, and energy savings [16–18]. This research gap underscores the need to clarify not only the chemical contribution of such sludge but also its impact on phase equilibrium, silicate and aluminate competition, and reaction kinetics during clinkerization.

In this context, the present study investigates the incorporation of semiconductor industrial sludge containing 11.4 wt.% fluorine into Portland cement clinker production. Raw mixes were prepared with 6%, 9%, and 12% replacement levels and sintered at temperatures ranging from 1300 °C to 1500 °C. The main objectives were to identify the optimal parameters for significant temperature reduction, evaluate the mineralogical and chemical transformations induced by sludge addition, and quantify the resulting energy savings and CO_2 emission reductions.

This study not only demonstrates the feasibility of valorizing semiconductor sludge as a novel mineralizer but also provides essential insights for potential industrial-scale applications and long-term performance assessment. The findings, which highlight a promising pathway to reduce the carbon footprint of cement manufacturing, are a significant step forward in the advancement of waste recycling and the development of eco-efficient and sustainable cement production technologies.

2. Materials and Methods

2.1. Raw Materials

In this study, reagent-grade calcium carbonate ($CaCO_3$, Junsei, extra pure), aluminum oxide (Al_2O_3), silicon dioxide (SiO_2), and iron (III) oxide (Fe_2O_3) (Samchun, extra pure) were employed as baseline raw materials for clinker preparation. In addition, a fluorine-rich semiconductor sludge, obtained from a local manufacturing facility in South Korea, was incorporated due to its elevated concentrations of calcium oxide (CaO, 36.8 wt.%), aluminum oxide (Al_2O_3, 19 wt.%), and fluorine (11.4 wt.%). The sludge was chemically characterized using inductively coupled plasma optical emission spectrometry (ICP-OES), carbon/sulfur analysis, and ion chromatography, with principal oxides expressed as weight percentages in Table 1. CaO and Al_2O_3 were expected to enhance the stability of silicate and aluminate phases, while the high fluorine concentration was anticipated to act as a mineralizing flux, lowering melting points and facilitating clinker phase formation. Ordinary Portland Cement (OPC), conforming to KSL 5201 (2013) and exhibiting a compressive strength of 42.5 MPa, was used as the reference material [19,20].

Table 1. Chemical composition of semiconductor waste sludge (wt.%).

Sample	SiO_2	CaO	P_2O_5	MgO	F	Fe_2O_3	Al_2O_3	C	SO_3	Others	LOI
Sludge (Sc)	3.84	36.8	3.29	3.16	11.4	0.36	19	4.39	3.88	>0.3	28.6

Note: LOI (Loss on Ignition) accounts for volatiles lost during heating; "Others" include trace elements below 0.3 wt.%.

2.2. Raw Mix Preparation and Chemical Modulus Calculation

Raw mixes were prepared by partially substituting clinker raw materials with semiconductor sludge at mass replacement levels of 6%, 9%, and 12%. The specific proportions of raw materials and sludge in each blend are provided in Table 2. These substitution levels were determined from preliminary trials and supporting literature: 6% were selected to examine the initial effect of sludge incorporation on clinker properties, whereas 9% and 12% represented incremental levels commonly employed in cement research. Substitutions beyond 12% were excluded to mitigate potential clinker quality deterioration and to avoid practical challenges during the sintering process.

Table 2. Raw mix design (g).

Sample	$CaCO_3$	Al_2O_3	SiO_2	Fe_2O_3	Sludge (Sc) %	(F) Content %
Reference	78.88	3.72	15.07	2.33	0	0
Blended 6	74.42	2.55	14.74	2.29	6	0.68
Blended 9	72.19	1.97	14.58	2.27	9	1.03
Blended 12	69.95	1.38	14.41	2.25	12	1.37

The schematic diagram in Figure 1 illustrates the process of incorporating semiconductor sludge into cement clinker production. In this process, fluorine-rich semiconductor waste is blended with conventional raw materials and then processed through the standard cement manufacturing steps. The presence of semiconductor sludge helps lower the required sintering temperature, which leads to reduced energy consumption and increased efficiency in clinker production. This approach not only offers energy savings but also promotes the utilization of industrial waste, contributing to more sustainable cement manufacturing practices.

To evaluate the suitability of the raw mixes for clinker formation, key chemical moduli were calculated based on oxide weight percentages:

- Lime Saturation Factor (LSF): Indicates the proportion of calcium oxide relative to silica, alumina, and iron oxide, and is critical for clinker burnability [21].

$$LSF = \frac{CaO}{2.8 \times SiO_2 + 1.2 \times Al_2O_3 + 0.65 \times Fe_2O_3} \times 100 \quad (1)$$

- Silica Modulus (SM): Ratio of silica to the combined alumina and ferric oxides, impacting the silicate phase formation [22].

$$SM = \frac{SiO_2}{Al_2O_3 + Fe_2O_3} \quad (2)$$

- Iron Modulus (IM): Ratio of ferric oxide to alumina, influencing aluminate and ferrite phase content [23].

$$IM = \frac{Fe_2O_3}{Al_2O_3} \quad (3)$$

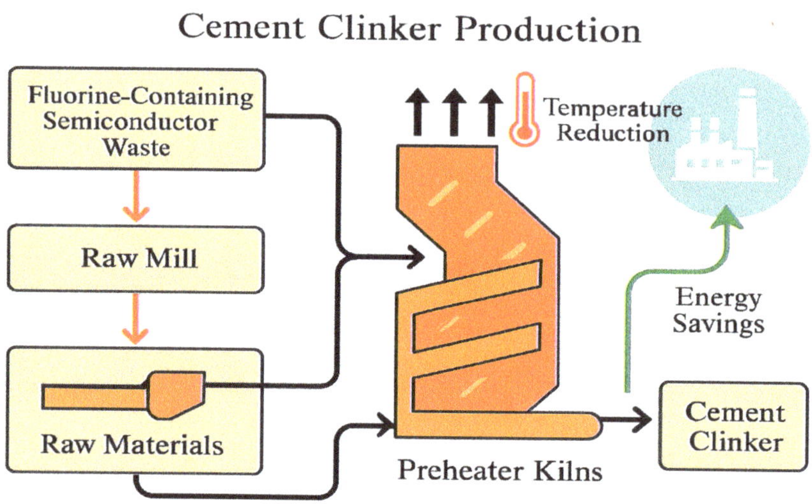

Figure 1. Clinker production process using semiconductor sludge.

The calculated moduli for all blended raw mixes were within optimal ranges (LSF ~92.0, SM ~2.5, IM ~1.6), indicating favorable conditions for clinker production.

2.3. Sample Preparation and Sintering Procedure

All raw materials were weighed according to the mix proportions specified in Table 2 to ensure sample consistency and reproducibility. The powders were homogenized in an HT-1000 ball mill for one hour using zirconia grinding media of 10 mm and 20 mm in diameter. The use of dual-sized media promoted both impact and attrition mechanisms, thereby enhancing particle size reduction and mixing uniformity.

Following milling, distilled water was gradually added at 35 wt.% of the total powder mass to achieve optimal plasticity for shaping [24]. The resulting paste was pressed manually into spherical pellets of ~25 g each, ensuring consistent size and geometry for uniform sintering behavior. The pellets were then dried at 120 °C for 24 h to eliminate residual moisture and prevent cracking or bloating during high-temperature treatment.

Dried spheres were sintered in a high-temperature electric furnace with an air-heating system, using a heating rate of ~10 °C/min. Sintering was conducted between 1300 °C and 1500 °C in 50 °C increments, after which the specimens were allowed to cool naturally to ~50–60 °C. Each sintering cycle was carefully controlled with respect to heating rate, dwell time, and furnace atmosphere to ensure reproducibility.

For comparison, reference spheres prepared from reference clinkers were subjected to the same sintering conditions, providing a baseline for evaluating the properties and performance of the sludge-incorporated samples.

Figure 2 presents photographs of cement clinker samples following the sintering process. The reference clinker sintered at 1300 °C displays a finer particle morphology, indicative of incomplete sintering. In contrast, the reference clinker processed at 1350 °C exhibits a coarser texture with larger and more aggregated particles, reflecting enhanced sintering progression. Notably, the blended clinker containing 12% semiconductor sludge and sintered at 1500 °C demonstrates a distinctly darker color and the presence of well-defined crystalline structures, suggesting significant phase development and possible alterations in mineralogical composition resulting from the higher temperature and sludge incorporation.

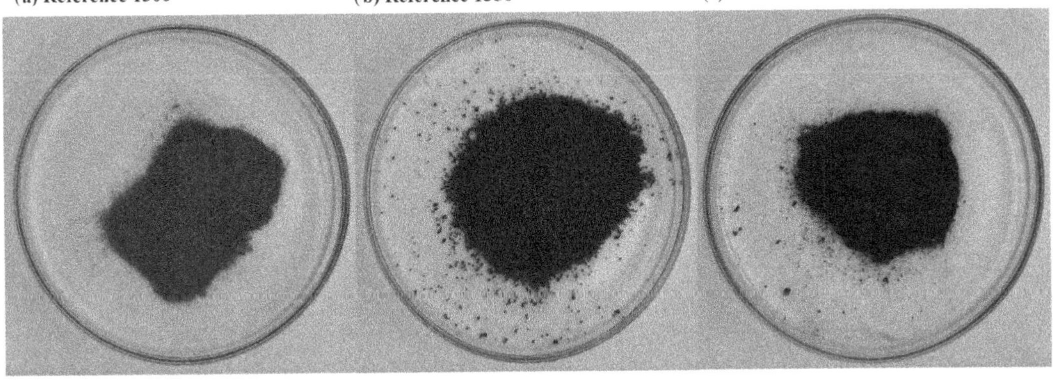

Figure 2. Photograph of dusted clinkers after sintering.

2.4. Analytical Methods

2.4.1. Determination of Free Calcium Oxide (FreeCaO)

The content of free calcium oxide, which indicates incomplete clinker formation, was determined using a chemical extraction method. Finely ground clinker was weighed with high precision and mixed with ethylene glycol, a solvent that selectively dissolves this phase. The suspension was incubated in a temperature-controlled water bath at 50–60 °C for about 30 min to ensure complete dissolution while minimizing interference from calcium silicate and carbonate phases. After incubation, the mixture was vacuum filtered to separate the undissolved solids, and the clear filtrate containing the dissolved fraction was carefully collected to avoid loss or contamination [25].

The filtrate was titrated with a standardized 0.1 M hydrochloric acid solution. Bromocresol green was used as the pH indicator because it exhibits a distinct color change around pH 4.7–5.2, allowing for the accurate determination of the titration endpoint. During titration, hydrochloric acid neutralizes the dissolved FreeCaO, and the color change from blue to yellow indicates the endpoint [26]. The volume of hydrochloric acid consumed at the endpoint was accurately recorded, and the FreeCaO content was calculated using the stoichiometric relationship between CaO and hydrochloric acid. The calculation accounted for the sample weight and dilution factors to express FreeCaO as a percentage of the clinker sample [27].

2.4.2. X-Ray Fluorescence (XRF) Spectroscopy

The major oxide compositions of clinker samples were quantitatively analyzed by X-Ray Fluorescence (XRF) spectroscopy using a RIGAKU Supermini 200 instrument (RIGAKU, Tokyo, Japan). To prepare the samples for analysis, powdered clinker was pressed into pellets and covered with polyethylene (P.E.) film to prevent contamination and sample loss while maintaining a consistent surface for measurement. This thin P.E. film is X-ray-transparent, allowing for accurate excitation and the detection of characteristic fluorescence without interfering with the analysis [28]. Each sample underwent measurement for 30 min to ensure high precision, accuracy, and reproducibility. The analysis targeted key oxides including calcium oxide (CaO), silicon dioxide (SiO_2), aluminum oxide (Al_2O_3), iron (III) oxide (Fe_2O_3), magnesium oxide (MgO), sulfur trioxide (SO_3), sodium oxide

(Na_2O), and potassium oxide (K_2O) [29]. The resulting quantitative data were used to verify the chemical composition of the clinker after sintering, confirming that the sintering process achieved the intended oxide proportions [30]. Furthermore, these measurements enabled a direct comparison with the raw mix proportions, providing insight into the consistency and efficiency of the clinker production process.

2.4.3. X-Ray Diffraction (XRD) Analysis

The mineralogical phase identification and quantification of clinker samples were systematically performed using a Bruker D6 Phaser X-ray diffractometer (Bruker, Karlsruhe, German), operated at an accelerating voltage of 40 kV and an emission current of 15 mA. This instrument configuration ensures sufficient X-ray intensity for clear diffraction signals and optimal resolution [31]. To prevent hydration and alteration of the clinker phases during sample preparation, samples were finely ground while suspended in 95% isopropanol [32]. The use of isopropanol as a grinding medium minimizes exposure to atmospheric moisture, thus preserving the original mineral phases for accurate analysis. The powdered samples were then evenly spread on a zero-background sample holder designed to reduce noise and enhance signal clarity [33,34].

Diffraction data were collected by scanning over a 2θ angular range from 5° to 70°, employing a step size typically between 0.02° and 0.05° 2θ, with a total scan time of approximately 3 h. This extended scan duration and acceptable angular resolution enable the acquisition of high-quality diffraction patterns with well-defined peaks, allowing for the precise identification of both major and minor mineralogical phases. For quantitative phase analysis, the Rietveld refinement technique was applied using HighScore Plus software version 3.0.5. The Rietveld method involves fitting a calculated diffraction pattern to the observed data by adjusting crystal structure parameters, peak shapes, background, and instrumental effects [35]. This comprehensive approach facilitates the accurate quantification of phase abundances, even in complex multiphase clinker samples. The resulting phase compositions provide critical insight into clinker quality, informing the optimization of raw material mixtures and sintering conditions to achieve desired cement performance characteristics.

2.5. Limitations and Future Work

This study demonstrated the potential of fluorine-rich semiconductor sludge as a mineralizer for reducing clinker sintering temperature; however, several limitations remain. Microstructural characterization techniques such as Scanning Electron Microscopy with Energy-Dispersive Spectroscopy (SEM/EDS), optical microscopy, and Fourier-Transform Infrared Spectroscopy (FTIR) were not employed in the present work. Incorporating optical microscopy in future investigations would allow for a more precise determination of phase size distribution and morphology, thereby complementing the crystallographic data obtained by XRD. Such analyses could provide deeper insights into the mechanisms by which fluorine affects phase formation and stabilization.

Moreover, this research was limited to clinker formation and phase evaluation under laboratory-scale sintering conditions. The hydration behavior, setting time, and mechanical properties of cementitious systems produced with the modified clinkers were not investigated. To establish the practical applicability of this approach, future studies will involve comprehensive mortar- and paste-level performance evaluations, including heat of hydration measurements, compressive strength testing, and durability assessments.

In addition, only natural cooling conditions were applied in this study. Exploring alternative cooling conditions, such as water quenching and controlled slow cooling, in future work will provide valuable insights into how cooling rate influences phase crystallinity, stability, and clinker reactivity.

Finally, the environmental and economic implications of large-scale sludge utilization must be considered. A life-cycle assessment (LCA) and cost–benefit analysis will be essential to validate the sustainability and industrial feasibility of replacing conventional raw materials with semiconductor sludge.

3. Results

3.1. FreeCaO Analysis

The FreeCaO content of clinker samples incorporating fluorine-containing semiconductor sludge at replacement levels of 6%, 9%, and 12% was systematically evaluated across sintering temperatures ranging from 1300 °C to 1500 °C. The results are summarized in Table 3.

Table 3. FreeCaO results.

Sample	1300 °C	1350 °C	1400 °C	1450 °C	1500 °C
Reference	62	8	1.6	0.4	0.4
Blended 6%	1.8	0.6	0.6	0.4	0.6
Blended 9%	0.4	0.6	0.4	0.4	0.4
Blended 12%	0.4	0.4	0.4	0.4	0.4

A detailed examination of Table 3 and Figure 3, together with the extended quantitative values presented in Appendix A (Table A1), reveals a pronounced difference in FreeCaO content between the reference OPC and the sludge-blended clinkers, highlighting the impact of fluorine-containing semiconductor sludge on clinker formation kinetics and burnability. At 1300 °C, the reference clinker exhibits a value of 62, indicating highly incomplete mineralization and insufficient solid-state reactions under these conditions. In stark contrast, the blends containing 9 and 12 units of semiconductor sludge show a drastic reduction to 0.4, providing compelling evidence that the fluorine-rich sludge acts as an effective mineralizer. This promotes raw mix reactivity and accelerates both decarbonation and silicate phase formation at lower temperatures.

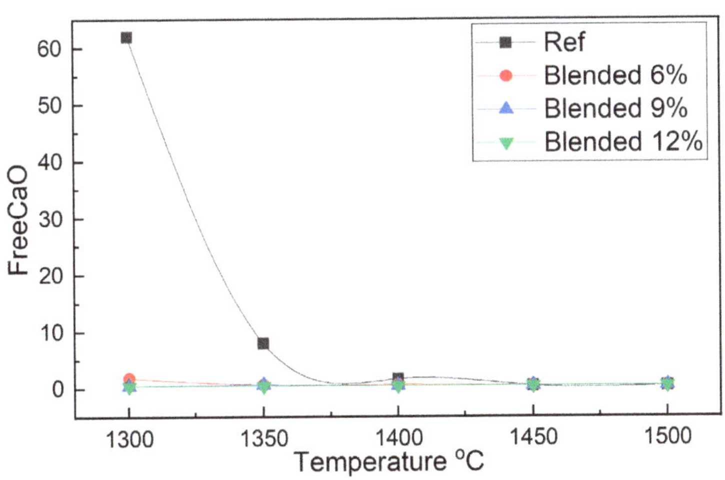

Figure 3. Burnability curves of reference and sludge-blended clinker samples showing FreeCaO content as a function of sintering temperature.

At 1350 °C, this beneficial effect is further corroborated: the reference clinker retains a relatively high value of 8, whereas the sludge-blended samples achieve near-complete conversion, with values of 0.6 and 0.4 for 9 and 12 units, respectively. The sharp reduction in residual lime in the sludge-containing clinkers across both temperatures suggests enhanced melt formation and greater ionic mobility, facilitating the incorporation of free lime into the principal clinker phases, particularly alite (C_3S) and belite (C_2S) [36].

These findings underscore the role of fluorine, introduced via semiconductor sludge, in reducing the activation energy required for key clinker-forming reactions. The fluxing action of fluorine lowers the melting point of the system, expands the temperature window for liquid phase formation, and promotes the earlier onset of mineralogical transformations. Consequently, complete clinker formation can be achieved at temperatures substantially lower than those in conventional practice without sacrificing product quality. Such improvements not only enhance thermal efficiency and reduce fuel consumption but also open new pathways for the sustainable utilization of industrial waste in cement manufacturing.

3.2. X-Ray Fluorescence Analysis

The chemical compositions of clinker samples prepared by partially replacing raw materials with fluorine-containing semiconductor sludge at 0%, 6%, 9%, and 12% were analyzed by XRF spectroscopy and compared to the reference clinker. This analysis quantified principal oxides, such as CaO, SiO_2, Al_2O_3, and Fe_2O_3, as well as minor oxides including MgO, SO_3, Na_2O, and K_2O, with high precision.

The comprehensive XRF analysis presented in Table 4 reveals systematic variations in clinker chemical composition as a function of semiconductor sludge replacement levels and sintering temperatures. The CaO content, a critical factor influencing clinker phase stability and hydraulic reactivity, remained consistently within the typical Portland cement clinker range (~60–68), demonstrating robust compositional integrity despite partial substitution of raw materials.

Table 4. XRF analysis results (wt.%).

Samples	CaO	SiO_2	Al_2O_3	Fe_2O_3	MgO	P_2O_5	SO_3	Cl	K_2O	SrO
Ref (1300)	66.60	26.20	2.57	3.7		0.8	0.01		0.06	0.03
Ref (1350)	67.00	20.90	4.66	3.98		0.98	0.01	0.01	0.57	0.04
Ref (1400)	67.50	19.20	5.45	4.19		0.95		0.01	0.06	0.04
Ref (1450)	68.52	20.83	5.77	3.89		0.84		0.01	0.06	0.04
Ref (1500)	67.10	20.30	5.41	4.12		0.88			0.06	0.04
B6 (1300)	68.89	19.69	5.51	3.88	0.59	1.16	0.15	0.01	0.05	0.04
B6 (1350)	68.57	19.70	5.92	3.81	0.51	1.99	0.15	0.01	0.05	0.04
B6 (1400)	68.08	19.44	6.30	4.18	0.65	1.12	0.08	0.01	0.07	0.04
B6 (1450)	68.44	19.82	6.11	3.96	0.34	1.09	0.10	0.01	0.05	0.04
B6 (1500)	68.56	19.79	6.03	3.78	0.41	1.16	0.13	0.01	0.05	0.04
B9 (1300)	67.37	19.80	6.39	4.04	0.63	1.30	0.32	0.01	0.07	0.04
B9 (1350)	67.57	20.39	6.03	3.94	0.32	1.35	0.26		0.07	0.04
B9 (1400)	66.97	20.66	6.30	3.92	0.45	1.22	0.30		0.05	0.04
B9 (1450)	67.62	19.92	6.16	4.19	0.43	1.23	0.27		0.06	0.04
B9 (1500)	67.33	19.96	6.22	4.35	0.51	1.23	0.27		0.06	0.04
B12 (1300)	66.12	21.00	6.36	3.80	0.72	1.36	0.42		0.06	0.03
B12 (1350)	67.14	20.90	5.73	3.99	0.33	1.37	0.39		0.07	0.04
B12 (1400)	66.80	20.72	5.95	4.22	0.39	1.39	0.40		0.06	0.04
B12 (1450)	67.24	20.12	5.84	4.38	0.52	1.36	0.38		0.07	0.04
B12 (1500)	67.26	21.25	4.89	4.09	0.41	1.46	0.46		0.06	0.04

Notably, the Al_2O_3 concentration increased proportionally with sludge addition, reflecting the high alumina content inherent to the semiconductor sludge. This chemical enrichment is expected to significantly influence clinker phase assemblage, particularly by promoting the formation and volume fraction of calcium aluminate phases (e.g., C_3A), which can affect cement hydration kinetics and early strength development.

Conversely, the SiO_2 content remained relatively stable across all samples, thereby maintaining the silica modulus within optimal limits essential for clinker quality. Minor oxides such as MgO, SO_3, Na_2O, and K_2O exhibited minimal fluctuations, indicating that sludge incorporation did not adversely impact these constituents. Notably, the fluorine-rich sludge appears to function as a flux during sintering, lowering the eutectic temperature and facilitating liquid phase formation. This enhances mass transport and accelerates phase transformations, consistent with the reductions in FreeCaO content shown in Figure 3 and the stabilized oxide concentrations illustrated in Figure 4. The fluxing effect of fluorine thus underpins the improved burnability and sintering efficiency observed in sludge-blended clinkers.

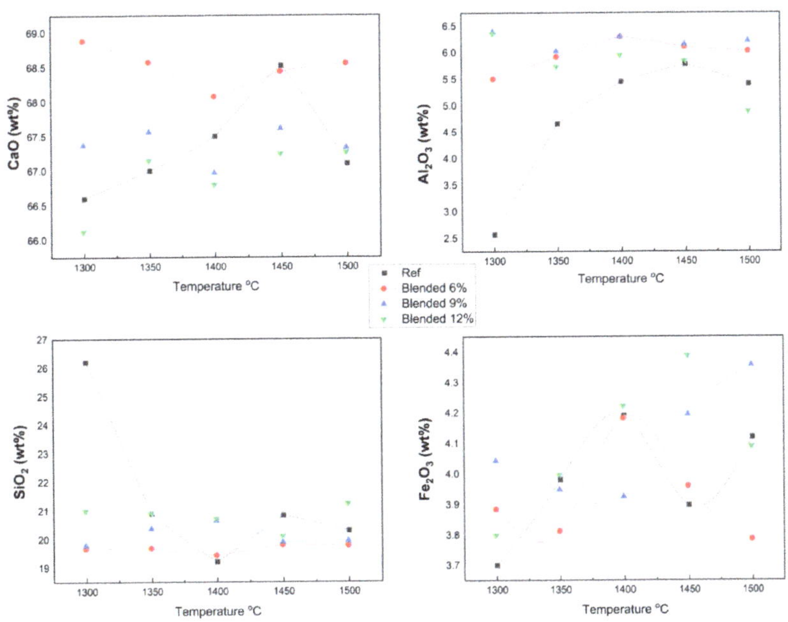

Figure 4. XRF results of principal oxides in clinker samples.

The calculated chemical indices, Lime Saturation Factor (LSF), Silica Modulus (SM), and Iron Modulus (IM) remain within optimal ranges for all blends, confirming the chemical viability for producing high-performance clinkers. Collectively, these results demonstrate that the partial replacement of raw materials with semiconductor sludge does not compromise clinker chemical integrity or phase equilibrium but rather introduces beneficial modifications enhancing sintering and clinker quality.

3.3. X-Ray Diffraction Analysis

The Clinker samples sintered between 1300 °C and 1500 °C with 6%, 9%, and 12% fluorine-containing semiconductor sludge were quantitatively analyzed by Rietveld refinement of XRD patterns. Figures 5 and 6 show detailed phase distributions, including alite polymorphs (C_3S M1, M3, T3, R), belite polymorphs (C_2S α, β, γ), ferrite (C_4AF), aluminate (C_3A), free lime, periclase, and a fluorine-containing compound ($Al_7Ca_6O_{16}F$).

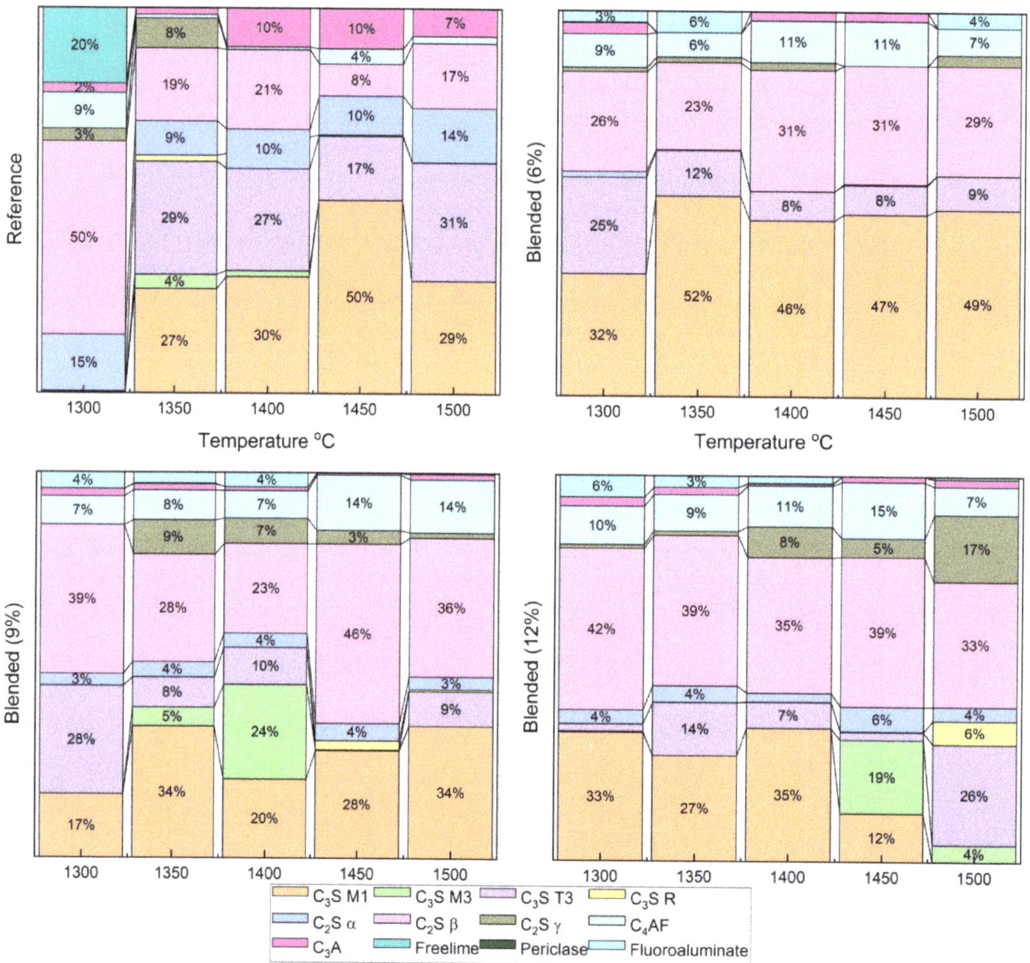

Figure 5. Rietveld refinement of XRD patterns for all clinker samples.

The reference clinker sintered at 1300 °C showed insufficient alite and belite formation, resulting in a poorly sintered, dusted structure, confirmed by high free lime content in Figures 3 and 5. In contrast, fluorine-rich sludge addition stabilized reactive belite polymorphs and altered alite polymorph distribution, increasing C_3S M3 and T3 phases.

These changes enable clinker formation at lower temperatures without compromising quality, enhancing sintering efficiency, and reducing cement production energy consumption. Reduced aluminate (C_3A) content and unique fluorine-bearing phases highlight sludge-derived fluorine's complex chemical role, potentially improving hydration kinetics and long-term durability.

(a) Fluorine-Induced Modulation of Alite (C_3S) Polymorph Distribution and Stability

The incorporation of fluorine-rich semiconductor sludge into the raw mix significantly alters the polymorphic distribution of alite phases. Quantitative analysis revealed a marked increase in metastable polymorphs C_3S M3 and T3 at the expense of the dominant stable C_3S M1 phase. This transformation suggests that fluorine acts to destabilize the equilibrium favoring C_3S M1, promoting dynamic phase transitions that enhance crystal nucleation and growth kinetics. Such modulation is critical as polymorphic form influences the hydration reaction rates and, consequently, the early strength development and durability of the cementitious matrix. The dynamic balance among alite polymorphs under fluorine influence likely facilitates more efficient hydration pathways and mechanical performance improvements.

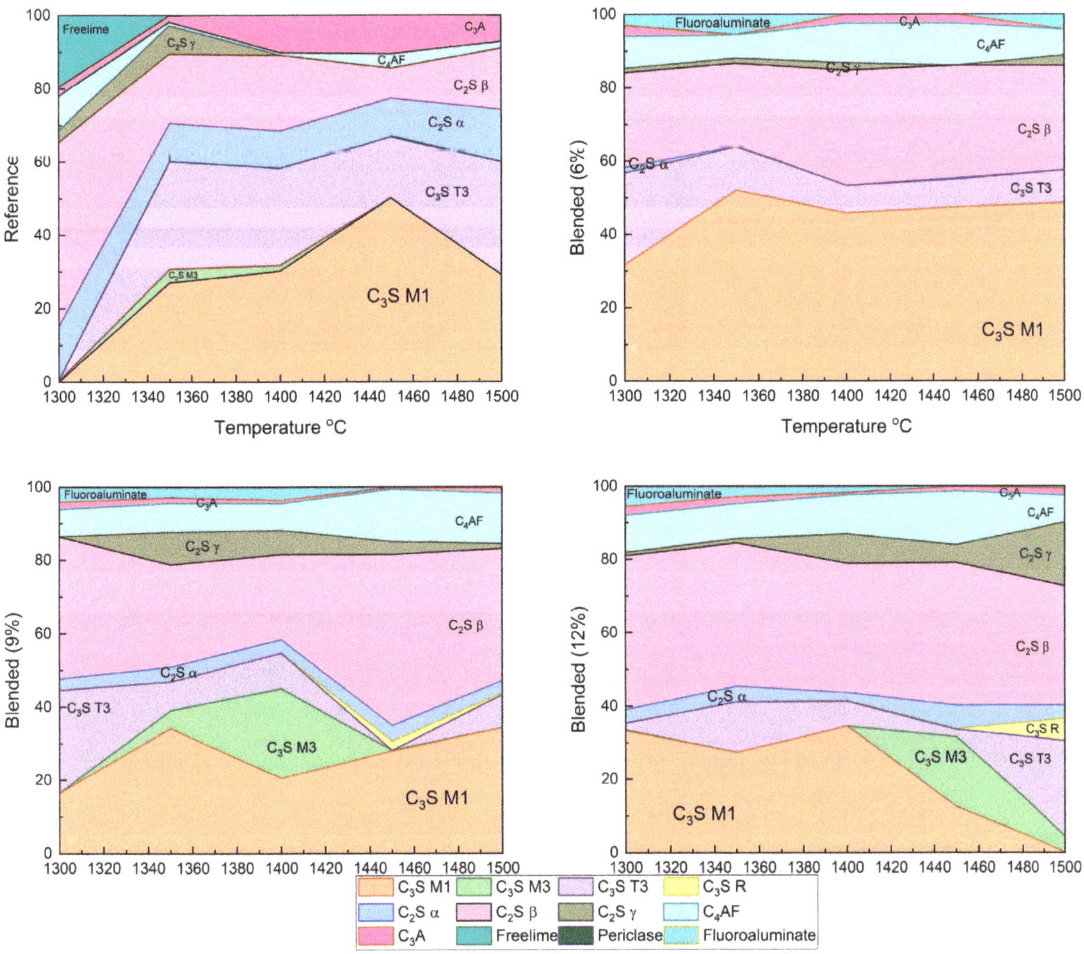

Figure 6. Rietveld refinement of XRD patterns showing phase composition of clinker samples with different sludge contents and sintering temperatures.

(b) Belite Enhanced Stabilization and Prevalence of Reactive Belite (C_2S β) Polymorph

The study found a consistent and significant increase in the highly reactive β-polymorph of belite (C_2S β) with increasing sludge content. This polymorph is known for its superior hydraulic reactivity and contribution to long-term strength gain in cement. Fluorine's stabilizing effect on this polymorph suggests an alteration of the clinker's phase equilibrium, enabling enhanced formation and retention of reactive belite at lower sintering temperatures. This stabilization not only supports improved mechanical properties over extended curing periods but also indicates a potential reduction in energy consumption by lowering the required sintering temperature for clinker production [37].

(c) The Impact on Aluminate (C_3A) and Ferrite (C_4AF) Phase Chemistry

Fluorine-rich sludge significantly suppressed the formation of the aluminate phase (C_3A), reducing its content to nearly a quarter of that observed in conventional OPC clinker. Since C_3A influences setting time and sulfate resistance, this suppression could lead to cement with enhanced durability and tailored setting characteristics. Concurrently, the ferrite phase (C_4AF) increased in concentration, which affects clinker color and hydraulic properties. The interplay of these phase shifts reflects complex chemical re-balancing induced by fluorine incorporation, which alters the overall clinker mineralogy and its functional properties.

(d) Free Near-Complete Reaction Evidenced by Minimal Free Lime (CaO) Content

The near absence of free lime across all samples, regardless of sludge content, indicates highly efficient sintering and reaction completeness. This observation confirms that fluorine addition does not compromise clinker quality but instead maintains or enhances the complete conversion of raw materials into stable clinker phases. The minimized free lime content is a crucial indicator of clinker stability and long-term performance.

(e) Formation of Unique Fluorine-Bearing Mineral Phase as Mineralizer

The exclusive detection of the $A_{17}Ca_6O_{16}F$ phase in sludge-containing clinkers substantiates the role of fluorine as a mineralizing agent. This unique fluorine-bearing phase integrates into the clinker lattice and promotes phase transformations at reduced sintering temperatures. By lowering the energy barrier for phase formation, this mineralizer facilitates clinker densification and phase homogeneity, ultimately improving clinker microstructure and mechanical integrity.

These fluorine-driven modifications of clinker mineralogy open pathways for energy-efficient cement production by enabling lower temperature sintering without sacrificing product quality. The altered phase assemblage enhances hydration kinetics and mechanical performance, particularly improving early strength gain and long-term durability. This study underscores the potential of using fluorine-containing industrial by-products, such as semiconductor sludge, as functional additives to tailor clinker chemistry, supporting sustainable manufacturing practices and advancing the development of high-performance cementitious materials.

Figure 7 shows powder X-ray diffraction patterns of blended clinkers B6 (black), B9 (red), and B12 (blue) sintered at 1450 °C, compared to a laboratory reference clinker (grey dashed). The major peaks correspond to belite (C_2S), with minor alite (C_3S) and a fluoride-stabilized calcium aluminate phase, $Ca_6Al_7O_{16}F$. Black arrows indicate $Ca_6Al_7O_{16}F$ reflections. The inset depicts the crystal structure of $Ca_6Al_7O_{16}F$ (Ca: blue; AlO_4 tetrahedra: grey; O/F: red). Intensities are normalized, and patterns are vertically offset for clarity.

3.4. Energy Saving and Carbon Emission Reduction Estimation

The incorporation of fluorine-containing semiconductor sludge into clinker production demonstrated the potential to reduce sintering temperature by approximately 100 to 150 °C compared to conventional Ordinary Portland Cement (OPC) clinkers, which typically sinter at around 1450 °C. This temperature reduction directly translates into significant energy savings and associated reductions in carbon dioxide (CO_2) emissions [38].

(a) Energy Consumption Reduction

Cement clinker production is an energy-intensive process, with fuel consumption strongly dependent on sintering temperature. Previous studies indicate that for every 10 °C decrease in sintering temperature, energy consumption decreases by approximately 1.5%. Applying this relationship to the observed temperature reduction yields an estimated energy saving of the following:

$$\Delta E = \frac{100 \sim 150}{10} \times 1.5\% = 15\% \sim 22.5\% \quad (4)$$

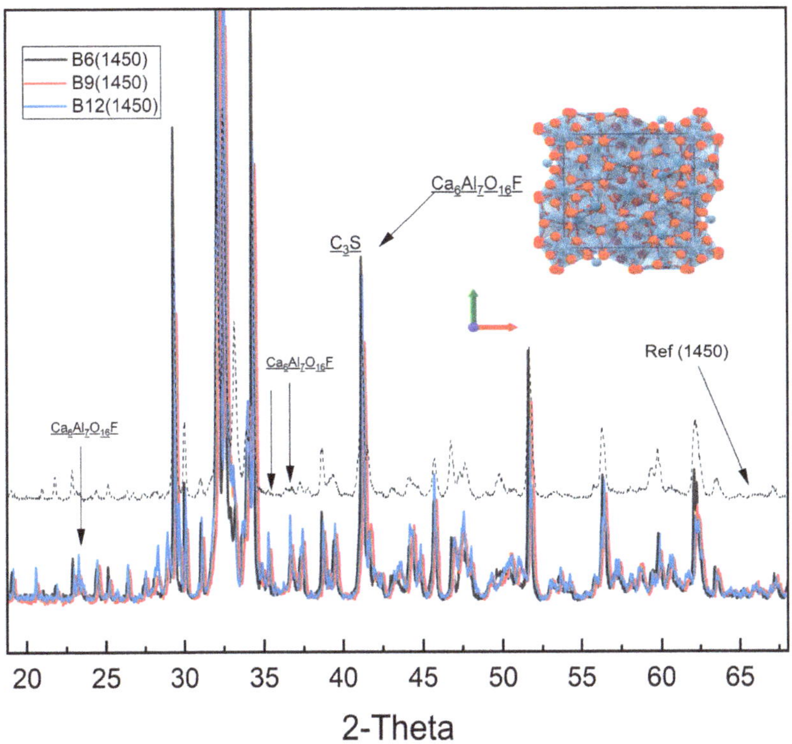

Figure 7. XRD patterns of fluoride-stabilized clinkers sintered at 1450 °C compared to reference clinker.

(b) Carbon Emission Reduction

Cement manufacturing accounts for approximately 8% of global anthropogenic CO_2 emissions, primarily due to the combustion of fossil fuels during the production of clinkers. Given that CO_2 emissions are roughly proportional to energy consumption, the reduction in emissions can be estimated as follows:

$$\Delta CO_2 = \Delta E \times \text{Current } CO_2 \text{ emissions per ton clinker} \quad (5)$$

As summarized in Table 5, a reduction in clinker sintering temperature by 100 to 150 °C has the potential to decrease energy consumption by approximately 15 to 22.5%. This substantial energy saving directly correlates with a significant reduction in carbon dioxide emissions, estimated to be between 0.105 and 0.203 tons of CO_2 per ton of clinker produced, assuming emission factors of 0.7 and 0.9 tons of CO_2 per ton of clinker.

The use of fluorine-rich semiconductor sludge as a mineralizer not only improves clinker quality and promotes phase formation at lower temperatures but also offers a pathway to reduce the environmental footprint of cement production. The potential energy savings and CO_2 emission reductions contribute to both economic and ecological sustainability, supporting global efforts to mitigate climate change.

Table 5. Energy savings and CO_2 reductions resulting from a decrease in clinker sintering temperature.

Temperature Reduction (°C)	Energy Savings (%)	CO_2 Emission Reduction (0.7 t CO_2/t Clinker)	CO_2 Emission Reduction (0.9 t CO_2/t Clinker)
100	15	0.105 t	0.135 t
150	22.5	0.158 t	0.203 t

4. Discussion

This study highlights the significant mineralogical and environmental benefits of incorporating fluorine-containing semiconductor sludge into cement clinker production. Integrating experimental results from XRD, XRF, and FreeCaO analyses provides a comprehensive understanding of fluorine's impact on clinker phase formation, sintering temperature reduction, and sustainability metrics.

(a) Mineralogical Impacts and Phase Transformations

Rietveld refinement of XRD data showed that fluorine alters the polymorphic distribution of alite (C_3S), favoring metastable phases C_3S M3 and T3. This polymorphic shift is likely to positively influence early hydration kinetics and mechanical strength developments [39]. The reactive belite (C_2S β) polymorph is stabilized and enriched with increasing sludge content, potentially improving long-term strength. Suppression of aluminate (C_3A) phases in sludge-containing clinkers suggests enhanced sulfate resistance and altered setting behavior [15]. The formation of fluorine-bearing phases such as $A_{17}Ca_6O_{16}F$ confirm fluorine's role as a mineralizer, facilitating clinker formation at lower temperatures [40]. Although the industrial sludge cannot be classified as a conventional pozzolan due to its limited amorphous silica content, its high CaO and Al_2O_3 concentrations actively contributed to clinker phase development, particularly favoring the formation of C_3S, C_2S, and C_3A. The elevated fluorine content further acted as a fluxing agent, lowering the burning temperature of the raw mix by ~150 °C and accelerating the crystallization of clinker minerals. These effects were corroborated by XRF and Rietveld-refined XRD analyses, which confirmed the presence of the expected silicate and aluminate phases.

At clinkerization temperatures, the sludge primarily decomposed into its oxide constituents (CaO and Al_2O_3), thereby influencing the chemical balance of the raw mix. While the excess CaO and Al_2O_3 could promote the formation of aluminate phases (C_3A and C_4AF) at the expense of silicates (C_3S and C_2S), the fluxing action of fluorine enhanced the melt phase and accelerated silicate crystallization. Consequently, clinker samples with 6%, 9%, and 12% sludge additions exhibited a stable phase assemblage in which silicate minerals remained dominant, demonstrating that the fluxing role of fluorine effectively mitigates the Ca–Al/Ca–Si competition.

(b) Environmental and Energy Implications

A reduction in clinker sintering temperature by approximately 100–150 °C is estimated to yield energy savings of about 15–22.5%. Considering the inherently high energy demand of clinker production, such a decrease translates into significant operational cost reductions as well as lower fossil fuel consumption [41]. Furthermore, based on reported emission factors of 0.7–0.9 tons of CO_2 per ton of clinker, this temperature reduction corresponds to a decrease of approximately 0.1–0.2 tons of CO_2 emissions per ton of clinker produced. This outcome highlights the potential of the proposed approach to mitigate the carbon footprint of the cement industry substantially [17].

Beyond direct reductions in fuel consumption and CO_2 emissions, lowering the sintering temperature also offers broader environmental benefits. Reduced thermal stress during the production process can minimize NO_x emissions, which are another major contributor to air pollution in cement manufacturing [42]. In addition, the partial substitution of conventional raw materials with semiconductor sludge promotes resource efficiency and advances circular economy practices by diverting industrial waste from landfills. This dual advantage reduction of greenhouse gas emissions and waste valorization further enhances the environmental sustainability of clinker production [43].

From an energy perspective, large-scale implementation of this approach could reduce the overall dependency of the cement industry on non-renewable fossil fuels. This is especially relevant under tightening environmental regulations and carbon taxation schemes, where energy efficiency directly translates into financial competitiveness. Future integration of renewable energy sources for kiln operations, combined with sintering temperature reduction strategies, may provide synergistic benefits in advancing the cement sector toward climate neutrality.

(c) Practical Considerations and Future Work

While the benefits of fluorine-containing sludge as a mineralizer are promising, further studies are required to evaluate the long-term durability of cements produced with modified clinker phases. Pilot- and industrial-scale investigations should be conducted to assess kiln operation stability, emission control, and product consistency, ensuring that the approach is technically and economically feasible under real production conditions.

Additionally, detailed research on phase transformation kinetics, hydration mechanisms, and the effects of different cooling regimes will provide deeper insight into the role of fluorine in clinker reactivity. Future work should also extend to evaluating hydration performance, mechanical strength, and durability, supported by life-cycle and cost–benefit analyses to confirm the environmental and industrial sustainability of this method.

(d) Sustainability Considerations of Fluorine Evaporation

During the thermal treatment of the sludge, some fluorine species may volatilize; however, under the experimental conditions of this study, the risk of direct HF emission is considered relatively low. This is because most of the fluorine tended to react with Ca and Al, forming stable crystalline phases such as CaF_2 or Ca–Al–F compounds, as confirmed by XRF and Rietveld-refined XRD analyses. Therefore, instead of being fully released into the atmosphere, the majority of fluorine remained incorporated into the clinker structure in a stable form.

In addition, fluorine acted as a fluxing agent, lowering the melting temperature of the raw mix and promoting the formation of clinker phases. As a result, the clinker burning temperature was reduced by approximately 150 °C, leading to decreased energy consumption and a significant reduction in CO_2 emissions. This represents an important contribution to the sustainability of cement production.

Hence, the overall sustainability of sludge recycling must be considered in terms of two aspects:

(a) Positive effect reduction of energy demand and CO_2 emissions
(b) Limited risk only a minor fraction of fluorine may volatilize during processing.

Our findings suggest that most of the fluorine remained bound within the clinker in the form of stable phases, and thus, the probability of HF release is relatively low. Nevertheless, future studies should include direct monitoring of possible gaseous emissions and evaluate emission control strategies (e.g., filtration or neutralization systems). In this way, the use of high-fluorine industrial sludge can combine the benefits of improved energy efficiency and CO_2 reduction while minimizing environmental risks.

5. Conclusions

1. This mineralogical transformation and performance benefits

Fluorine-containing semiconductor sludge, when used as a partial substitute in clinker production, stabilizes the reactive β-polymorph of belite, alters the polymorphic distribution of alite, and promotes the formation of distinct fluorine-bearing phases. These mineralogical changes accelerate early hydration kinetics and improve the mechanical strength development of cementitious materials, confirming the performance advantages of sludge incorporation.

2. Energy and environmental implications

The incorporation of fluorine enables a reduction in clinker sintering temperature of up to 150 °C without compromising phase quality or stability. This thermal decrease results in estimated energy savings of 15–22.5% and a reduction of approximately 0.1–0.2 tons of CO_2 emissions per ton of clinker, thereby providing clear environmental and operational benefits.

3. Pathway for sustainable cement production

Collectively, the findings demonstrate a promising approach to sustainable cement manufacturing through the valorization of industrial semiconductor waste. Future work should prioritize scale-up studies, long-term durability assessments, and mechanistic investigations of fluorine's influence on hydration and phase transformation to optimize clinker formulations and advance eco-efficient cement technologies.

Author Contributions: Conceptualization, B.M. and Y.-S.C.; methodology, B.M.; software, B.M. and Y.L.; validation, B.M., Y.L. and Y.-S.C.; formal analysis, B.M.; investigation, B.M. and Y.-S.C.; resources, Y.L. and Y.-S.C.; data curation, B.M.; writing—original draft preparation, B.M.; writing—review and editing, Y.L., J.-H.J.K. and B.M.; supervision, J.-H.J.K. and Y.-S.C.; funding acquisition, Y.-S.C.; correspondence, J.-H.J.K. and Y.-S.C.; additional contributions including project coordination, technical guidance, and manuscript revision support, Y.-S.C. All authors have read and agreed to the published version of the manuscript.

Funding: This work was supported by the Technology Innovation Program (RS-2024-00438915, the development of technology for the continuous process of low-temperature sintering clinkers and the high efficiency of preheating and cooling processes) funded by the Ministry of Trade, Industry, and Energy (MOTIE, Korea).

Data Availability Statement: The original contributions presented in this study are included in the article. Further inquiries can be directed to the corresponding author.

Acknowledgments: We sincerely thank SAMPYO Cement Company and their dedicated R&D team for their generous support and valuable contributions to this study. Their willingness to share critical information was instrumental in the successful completion of our research. We deeply appreciate their collaboration, which greatly enriched the depth and quality of this work.

Conflicts of Interest: The authors declare no conflicts of interest.

Abbreviations

XRD	X-Ray Diffraction
XRF	X-Ray Fluorescence
OPC	Ordinary Portland Cement
FreeCaO	Free Calcium Oxide
C_3S	Tricalcium Silicate (Alite)
C_2S	Dicalcium Silicate (Belite)
C_3A	Tricalcium Aluminate
C_4AF	Tetracalcium Aluminoferrite (Ferrite)
M1, M3, T3, R	Polymorphs of Alite (C_3S)

Appendix A

This appendix contains supplementary experimental data and detailed calculations supporting the main findings presented in the manuscript.

- XRD Rietveld Refinement Data: Complete phase quantification results for all clinker samples at various sintering temperatures, including minor phases in the main text.
- FreeCaO Measurements: Detailed raw data and calculation methods for free calcium oxide content across samples.
- XRF Elemental Analysis: Full elemental composition tables with raw spectral data used for Thermo-Calc input.
- Energy and CO_2 Emission Calculations: Step-by-step calculation sheets for estimating energy savings and carbon emission reductions based on sintering temperature reductions, including assumptions and formula derivations.
- Experimental Setup: Photographs and schematic diagrams of the experimental setup and sample preparation protocols.

All raw data and calculation spreadsheets are available from the authors upon reasonable request.

Table A1. Extended XRD phase quantification data.

Sample ID	C_3S M1	C_3S M3	C_3S T3	C_2S b	C_4AF	C_3A	FreeCaO	Other Phases
Ref-1450 °C	49.4	0	16.3	8.1	3.9	10.3	0	12.0
B6-1300 °C	31.7	0	25.1	25.9	8.7	2.9	0.1	11.6

References

1. Jessa, E.; Ajidahun, A. Sustainable practices in cement and concrete production: Reducing CO_2 emissions and enhancing carbon sequestration. *World J. Adv. Res. Rev.* **2024**, *22*, 2301–2310. [CrossRef]
2. Luo, W.; Shi, Z.; Wang, W.; Yu, A.; Liu, Z. Eco-Friendly Cement Clinker Manufacturing Using Industrial Solid Waste and Natural Idle Resources as Alternative Raw Materials. *Energy Environ. Manag.* **2025**, *1*, 59–65. [CrossRef]
3. Vasiliu, L.; Lazar, L.; Rusu, L.; Harja, M. Effect of Alternative Raw Materials Over Clincker Quality and CO_2 Emissions. *Sci. Study Res. Chem. Chem. Eng. Biotechnol. Food Ind.* **2024**, *25*, 325–334. [CrossRef]
4. Liu, L. Carbon Emission Assessment and Environmental Impact of Cement in the Context of Carbon Neutrality. *Appl. Comput. Eng.* **2025**, *159*, 77–84. [CrossRef]
5. Abdul-Wahab, S.A.; Al-Dhamri, H.; Ram, G.; Chatterjee, V.P. An overview of alternative raw materials used in cement and clinker manufacturing. *Int. J. Sustain. Eng.* **2021**, *14*, 743–760. [CrossRef]
6. Tamayo, P.; del Angel, G.G.; Setién, J.; Soto, A.; Thomas, C. Feasibility of silicomanganese slag as cementitious material and as aggregate for concrete. *Constr. Build. Mater.* **2023**, *364*, 129938. [CrossRef]
7. Flegar, M.; Serdar, M.; Londono-Zuluaga, D.; Scrivener, K. Regional Waste Streams as Potential Raw Materials for Immediate Implementation in Cement Production. *Materials* **2020**, *13*, 5456. [CrossRef]

8. Mangi, S.A.; Memon, Z.A.; Khahro, S.H.; Memon, R.A.; Memon, A.H. Potentiality of Industrial Waste as Supplementary Cementitious Material in Concrete Production. *Int. Rev. Civ. Eng. (IRECE)* **2020**, *11*, 214–221. [CrossRef]
9. Kırgız, M.S.; Khatib, J.; de Sousa Galdino, A.G.; Kenai, S.; Ashteyat, A.; Campilho, R.D.S.G.; Gonzales Lezcano, R.A.; Bamigboye, G.O.; Nagaprasad, N.; Ramaswamy, K. Waste marble sludge and calcined clay brick powders in conventional cement farina production for cleaner built environment. *Sci. Rep.* **2025**, *15*, 3467. [CrossRef]
10. Shyamala, G.; Hemalatha, B.; Gunavathie, M.A.; Ezhil, S.S.; Sathya, V.; Gobinath, M.; Babu, M.D. Evaluation of Paper Industry waste sludge and nano materials on properties of concrete. *E3S Web Conf.* **2025**, *621*, 01018. [CrossRef]
11. Jang, C.; Abebe, T.N. Utilizing Wheel Washing Machine Sludge as a Cement Substitute in Repair Mortar: An Experimental Investigation into Material Characteristics. *Materials* **2024**, *17*, 2037. [CrossRef] [PubMed]
12. Rusănescu, C.O.; Voicu, G.; Paraschiv, G.; Begea, M.; Purdea, L.; Petre, I.C.; Stoian, E.V. Recovery of Sewage Sludge in the Cement Industry. *Energies* **2022**, *15*, 2664. [CrossRef]
13. Wang, D.; Zhang, Q.; Liu, B.; Zhu, D.; Chen, Q. Enhanced immobilization of fluoride in phosphogypsum-based cement-free paste backfill modified by polyaluminum chloride and its mechanism. *Constr. Build. Mater.* **2025**, *458*, 139622. [CrossRef]
14. Li, Q.; Gu, S.; Yu, L.; Han, F.; Dong, J.; Li, C.; Jia, J. Fluorine Fixation Mechanisms of CaF_2 and $Ca_4Si_2O_7F_2$ during Spent Potlining Incineration with $CaSiO_3$ for Zero-Waste Utilization of Steel Waste Heat. *Steel Res. Int.* **2025**, 2500225. [CrossRef]
15. Da, Y.; He, T.; Ma, X.; Yang, R.; Lin, Y.; Yang, Q.; Bai, C. Probing into the Triggering Effects of Zinc Presence on the Mineral Formation, Hydration Evolution, and Mechanical Properties of Fluorine-Bearing Clinker. *Langmuir* **2025**, *41*, 11417–11427. [CrossRef]
16. Sinharoy, A.; Lee, G.Y.; Chung, C.M. Optimization of Calcium Fluoride Crystallization Process for Treatment of High-Concentration Fluoride-Containing Semiconductor Industry Wastewater. *Int. J. Mol. Sci.* **2024**, *25*, 3960. [CrossRef]
17. Bouregba, A.; Diouri, A.; Elghattas, B.; Boukhari, A.; Guedira, T. Influence of Fluorine on Clinker burnability and mechanical properties of CPA Moroccan cement. *MATEC Web Conf.* **2018**, *149*, 01075. [CrossRef]
18. Novosyolov, A.; Yu, D.; Novoselova, I.; Yu, V. Efficiency of Using a Technogenic Product of Electrolytic Production of Aluminum as a Mineralizer When Firing Portland Cement Clinker. *Bull. Belgorod State Technol. Univ. Named After V. G. Shukhov* **2022**, *7*, 71–80. [CrossRef]
19. Mend, B.; Lee, Y.J.; Kwon, D.Y.; Bang, J.H.; Chu, Y.S. Utilisation of industrial sludge-derived ferrous sulfate for hexavalent chromium mitigation in cement. *Adv. Cem. Res.* 2025; *in press*. [CrossRef]
20. Novosyolov, A.; Yu, D.; Novoselova, I.; Yu, V.; Krapivin, S. Study of The Mineralizing Effect of Cryolite and Its Influence on the Processes of Clinker Formation. *Bull. Belgorod State Technol. Univ. Named After V. G. Shukhov* **2025**, *10*, 86–95. [CrossRef]
21. Liu, Y.; Cui, Y.; Tong, X. The regeneration of cement from completely recyclable mortar: Effect of raw materials compositions. *Adv. Cem. Res.* **2024**, *37*, 253–268. [CrossRef]
22. Chu, Y.-S.; Lee, Y.-J.; Kim, N.-I.; Cho, J.-H.; Seo, S.-K. A Study on the Characteristics of Clinker and Cement as Chlorine Content. *Resour. Recycl.* **2021**, *30*, 10–16. [CrossRef]
23. Guo, X.; Wang, S.; Lu, L.; Wang, H. Influence of barium oxide on the composition and performance of alite-rich Portland cement. *Adv. Cem. Res.* **2012**, *24*, 139–144. [CrossRef]
24. Pereira, M.M.L.; de Souza, A.L.R.; Capuzzo, V.M.S.; de Melo Lameiras, R. Effect of the water/binder ratio on the hydration process of Portland cement pastes with silica fume and metakaolin. *Rev. IBRACON Estrut. Mater.* **2021**, *15*, e15105. [CrossRef]
25. Lau, O.W.; Luk, S.F.; Cheng, N.L.N.; Woo, H.Y. Determination of free lime in clinker and cement by Iodometry: An undergraduate experiment in redox titrimetry. *J. Chem. Educ.* **2001**, *78*, 1671–1673. [CrossRef]
26. Tobón, J.I.; Díaz-Burbano, M.F.; Restrepo-Baena, O.J. Optimal fluorite/gypsum mineralizer ratio in Portland cement clinkering. *Mater. Constr.* **2016**, *66*, e086. [CrossRef]
27. Mao, C.; Zhang, W.; Zhang, L.; Ma, Z.; Cui, Q.; Wang, H. Research on Preparation Process of Lithium Slag for High Belite Sulfoaluminate Cement Clinker. *Waste Biomass Valoriz.* **2025**, 1–11. [CrossRef]
28. Khelifi, S.; Ayari, F.; Tiss, H.; Chehimi, D.B.H. X-ray fluorescence analysis of Portland cement and clinker for major and trace elements: Accuracy and precision. *J. Aust. Ceram. Soc.* **2017**, *53*, 743–749. [CrossRef]
29. *ASTM C114-18*; Test Methods for Chemical Analysis of Hydraulic Cement. ASTM: West Conshohocken, PA, USA, 2018. [CrossRef]
30. Abdul, W.; Rößler, C.; Kletti, H.; Mawalala, C.; Pisch, A.; Bannerman, M.N.; Hanein, T. On the variability of industrial Portland cement clinker: Microstructural characterisation and the fate of chemical elements. *Cem. Concr. Res.* **2025**, *189*, 107773. [CrossRef]
31. Tlamsamania, D.; Ait-Mouha, M.; Slassi, S.; Khaddam, Y.; Zuluaga, D.L.; Yamni, K. Quantitative phase analysis of anhydrous clinker Portland using Rietveld method. *Rev. Inorg. Chem.* **2023**, *43*, 189–199. [CrossRef]
32. Santos, R.L.; Horta, R.B.; Pereira, J.; Nunes, T.G.; Rocha, R.; Canongia Lopes, J.N.; Colaço, R. Novel high-resistance clinkers with $1.10 < CaO/SiO_2 < 1.25$: Production route and preliminary hydration characterization. *Cem. Concr. Res.* **2016**, *85*, 39–47. [CrossRef]

33. Her, S.; Park, J.; Li, P.; Bae, S. Feasibility study on utilization of pulverized eggshell waste as an alternative to limestone in raw materials for Portland cement clinker production. *Constr. Build. Mater.* **2022**, *324*, 126589. [CrossRef]
34. Cao, F.; Ye, J.; Sun, Z.; Ren, X.; Yao, X.; Zhai, M. Compressive strength prediction of Portland cement clinker using machine learning by XRD data. *J. Sustain. Cem. Based Mater.* **2024**, *13*, 1120–1131. [CrossRef]
35. Kim, H.; Yoon, J. Automated Mineral Identification of Pozzolanic Materials Using XRD Patterns. *J. Mater. Civ. Eng.* **2024**, *36*, 04024415. [CrossRef]
36. Reno, M.L.G.; Silva, R.J.; Melo, M.L.N.M.; Boas, S.B.V. An Overview of Industrial Wastes as Fuel and Mineralizer in the Cement Industry. *Lat. Am. Appl. Res. Int. J.* **2016**, *46*, 43–50. [CrossRef]
37. Zhao, S.; Liu, Z.; Wang, F.; Hu, S. Elucidating factors on the strength of carbonated compacts: Insights from the carbonation of γ-C2S, β-C2S and C3S. *Cem. Concr. Compos.* **2025**, *155*, 105806. [CrossRef]
38. He, T.; Da, Y.; Shi, C.; Wang, S.; Wang, J.; Feng, Y.; Li, Z. Recovery of thermally treated fluorine-containing sludge as the substitutions of Portland cement. *J. Clean. Prod.* **2020**, *260*, 121030. [CrossRef]
39. Isteri, V.; Ohenoja, K.; Hanein, T.; Kinoshita, H.; Illikainen, M.; Tanskanen, P.; Fabritius, T. The Effect of Fluoride and Iron Content on the Clinkering of Alite-Ye'elimite-Ferrite (AYF) Cement Systems. *Front. Built Env.* **2021**, *7*, 698830. [CrossRef]
40. Kacimi, L.; Simon Masseron, A.; Ghomari, A.; Derriche, Z. Reduction of clinkerisation temperature by addition of salts containing fluorine. *Can. J. Civ. Eng.* **2006**, *33*, 1090–1097. [CrossRef]
41. Toure, A.O.; Auyeung, N.; Sambe, F.M.; Malace, J.S.; Lei, F. Cement Clinker based on industrial waste materials. *J. Civ. Environ. Eng.* **2020**, *10*, 1–5. [CrossRef]
42. Lin, H.; Yao, X.; Wang, X.; Zhang, X.; Mao, H.; Zhang, J.; Li, X.; Shu, Y.; Vujanovis, M.; Mikulčić, H. Green ammonia co-firing in cement rotary kiln: Numerical investigation on clinker quality and NOx emission by gas–solid interaction. *Appl. Therm. Eng.* **2025**, *277*, 127116. [CrossRef]
43. Verma, Y.K.; Ghime, D.; Mazumdar, B.; Ghosh, P. Emission reduction through process integration and exploration of alternatives for sustainable clinker manufacturing. *Int. J. Environ. Sci. Technol.* **2023**, *20*, 13329–13346. [CrossRef]

materials

MDPI

Article

The Governing Role of Si/Al Ratio in the Structural Evolution and Mechanical Properties of N-A-S-H Gel

Min Hu [1,2,*], Jiayun Chen [1,2], Bo Xia [3] and Jiejin Chen [1,2,*]

[1] School of Civil and Environmental Engineering, Changsha University of Science and Technology, Changsha 410114, China; chenjiayun@csust.edu.cn

[2] Hunan Provincial Engineering Research Center for Bridge Engineering Safety Control Technology and Equipment, Changsha 410114, China

[3] School of Hydraulic and Ocean Engineering, Changsha University of Science and Technology, Changsha 410114, China; boxia@csust.edu.cn

* Correspondence: hum07@lzu.edu.cn (M.H.); chenjj@csust.edu.cn (J.C.)

Abstract

Alkali-activated cementitious materials are environmentally friendly alternatives to traditional cement. The structure of their core product, sodium aluminosilicate hydrate (N-A-S-H) gel, is regulated by the silicon-to-aluminum (Si/Al) ratio; however, the atomic-scale mechanism underlying this influence remains unclear. Integrating reactive force field molecular dynamics simulations and experiments, this study systematically reveals the regulation mechanism of the Si/Al ratio (1.0–2.0) on the microstructure and macroscopic properties of N-A-S-H gels. Starting from well-defined PS and PSS oligomers, the simulation results demonstrate that the Si/Al ratio governs the polymerization pathway, aluminum coordination environment (especially the content of pentacoordinate aluminum), and evolution of nanoporosity. When the Si/Al ratio is approximately 1.8, the system exhibits the highest silicate polymerization degree, lowest nanoporosity, and densest three-dimensional (3D) network structure; deviation from this ratio leads to structural degradation due to charge imbalance or excessive polymerization. These computational findings are validated by experiments on fly ash-based geopolymers: the material achieves the highest compressive strength at a Si/Al ratio of 1.8. The consistency between simulations and experiments collectively reveals a cross-scale action mechanism: the Si/Al ratio determines the macroscopic mechanical properties by regulating the nanoscale packing density and defect distribution of the gel. This study provides critical atomic-scale insights for the rational design of high-performance geopolymers.

Keywords: alkali-activated cementitious materials; molecular dynamics; low-calcium fly ash; N-A-S-H gel; reaxff

1. Introduction

Geopolymers, as inorganic polymeric materials formed by dissolution-condensation reactions of alkali-activated aluminosilicate precursors (e.g., low-calcium fly ash, metakaolin), are widely regarded as one of the most promising sustainable cementitious materials to replace traditional Portland cement due to their excellent mechanical properties, outstanding durability, and significantly reduced carbon emissions [1,2]. Their core performance stems from the sodium aluminosilicate hydrate (N-A-S-H) gel as the binding phase, which is a three-dimensional amorphous network structure composed of $[SiO_4]^{4-}$ and $[AlO_4]^{5-}$ tetrahedra connected by shared oxygen atoms [3,4]. However, N-A-S-H gel is inherently

https://doi.org/10.3390/ma19020246

amorphous with long-range disorder, and the structure-activity relationship between its microstructure (e.g., local order, aluminum coordination environment, nanoscale pores) and macroscopic properties remains a key scientific bottleneck restricting the rational design and performance optimization of geopolymer materials [5].

Among numerous regulatory factors, the molar silica-alumina ratio (Si/Al) of the system is a core parameter affecting the formation and final performance of N-A-S-H gel, directly determining the setting time, mechanical strength, and durability of the material [6]. Numerous experiments have shown that macroscopic properties such as compressive strength first increase and then decrease with the change in Si/Al ratio, with an optimal value in the range of 1.5–2.0 [7–10]. Although characterization techniques such as X-ray diffraction (XRD), Fourier transform infrared spectroscopy (FTIR), scanning electron microscopy (SEM), and mercury intrusion porosimetry (MIP) can provide information on phase composition, chemical bonds, microstructure, and pore structure [11–13], it is difficult to dynamically capture the details of structural evolution at the atomic/nanoscale, such as the distribution of aluminum coordination states, condensation reaction pathways, and the formation process of nanoscale pores. Therefore, the microscopic mechanism by which Si/Al affects macroscopic properties remains unclear.

With the development of reactive force fields (ReaxFF), molecular dynamics (MD) simulation has become an important means to explore the gel formation mechanism at the atomic scale. It can effectively simulate chemical reaction processes such as hydrolysis and condensation [14–18], and has been used to investigate the influence of Si/Al ratio on the microstructure [11], thermal conductivity [19], and mechanical properties [20,21] of gels. However, most existing simulations start from "averaged" amorphous macromolecules or oligomers, failing to dynamically simulate their assembly process in competitive condensation based on well-defined structural units (e.g., PS, PSS). Thus, it is difficult to understand how Si/Al regulates network topology from the source of the reaction, and it is impossible to accurately explain the correlation mechanism between microstructure evolution and macroscopic properties.

The structural unit model proposed by Davidovits (e.g., PS, PSS) provides a classic framework for describing the short-range order of gels [22]. In recent years, advanced characterization techniques such as solid-state nuclear magnetic resonance (NMR) have provided experimental evidence for verifying the evolution of different structural units with Si/Al ratio [23,24], and SEM-EDS, XRD, and FTIR characterizations have also confirmed the influence of Si/Al ratio on the formation of structural units [25]. These advances indicate that starting from basic units with clear chemical definitions such as PS and PSS, adopting a bottom-up simulation approach to dynamically track their condensation and assembly processes will be more conducive to revealing the regulatory mechanism of Si/Al ratio on the structure of N-A-S-H gel.

To this end, this study designed a closed-loop research combining molecular dynamics simulation and multi-scale experimental verification. In terms of simulation, the validated ReaxFF reactive force field was adopted, with stoichiometrically precise PS (Si/Al = 1) and PSS (Si/Al = 2) as initial species to dynamically simulate their competitive condensation and gelation processes, so as to reveal the influence mechanism of Si/Al ratio on network atomic packing, connection topology, and nanoscale pores. In terms of experiments, alkali-activated low-calcium fly ash specimens with different Si/Al ratios were prepared by adjusting the ratio of glass powder to sodium silicate, and characterized by SEM, XRD, and mechanical tests. Based on the analysis of structural evolution at the atomic scale, this study aims to clarify the microscopic mechanism by which Si/Al ratio regulates the formation and performance optimization of N-A-S-H gel, and provide theoretical and data support for the rational design and performance improvement of high-performance geopolymers.

 https://doi.org/10.3390/ma19020246

2. Materials and Methods

2.1. Molecular Dynamics Simulation

2.1.1. Simulation Strategy and Model Construction

Based on the structural unit model proposed by Davidovits [22], this study adopted a bottom-up strategy to construct gel models. Polyaluminosilicate (PS, $SiAlO_7H_6$, Si/Al = 1) and polyaluminosiloxane (PSS, $Si_2AlO_{10}H_8$, Si/Al = 2) oligomers were employed as the basic building units, and examples of PS and PSS oligomers are shown in Figure 1a and Figure 1b, respectively. By precisely controlling the quantity ratio of the two oligomers, four initial systems with target Si/Al ratios (1.0, 1.5, 1.8, 2.0) were established in cubic periodic boundary boxes (side length: 32.5 Å) using Materials Studio 2020 (20.1.0.2728). A corresponding number of Na^+ ions (from NaOH) was added to each system to balance the charge of $[AlO_4]^{5-}$ tetrahedra generated from oligomer hydrolysis, and sufficient water molecules were introduced (with a mass ratio of water to cementitious materials of 0.4:1) to simulate a high-alkalinity solution environment, ensuring the full dissolution and migration of reactants. The four constructed model systems (with Si/Al ratios of 1.0, 1.5, 1.8 and 2.0, respectively) had initial densities ranging from 1.44 to 1.97 g/cm^3, which increased systematically with the increase in the Si/Al ratio and is consistent with the actual concentration range of alkali-activated pastes. The detailed initial atomic compositions, box sizes and key parameters are listed in Table 1.

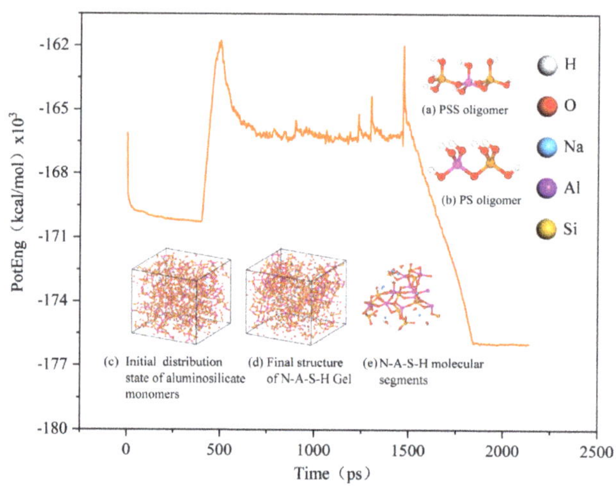

Figure 1. Variation in system potential energy during simulation, Network Structure of N-A-S-H Gel and PS, PSS.

Table 1. Initial Compositions and Parameters of the Molecular Dynamics Simulation System.

System ID	Target Si/Al Ratio	Number of PS	Number of PSS	Number of Na^+ (NaOH)	Initial H_2O Molecules	Final H_2O Molecules	Total Density (g/cm^3)
GP-1.0	1	100	0	100	473	694	1.443
GP-1.5	1.5	50	50	100	560	807	1.708
GP-1.8	1.8	20	80	100	612	848	1.867
GP-2.0	2	0	100	100	647	898	1.973

https://doi.org/10.3390/ma19020246

2.1.2. Force Field Selection and Simulation Details

The Reactive Force Field (ReaxFF) developed by van Duin et al. [26,27] was adopted for simulations. Based on the bond order concept, this force field can dynamically describe the formation and cleavage of chemical bonds, making it suitable for simulating condensation reactions in alkali-activated environments. Its potential energy function incorporates bond order-dependent bond energy, bond angle, dihedral angle, van der Waals, and Coulomb interaction terms. The ReaxFF parameters for Na, Si, Al, O, and H were derived from previously published results [27–30], whose reliability has been fully validated.

Molecular dynamics simulations were performed using the LAMMPS software (4 February 2025) with a time step set to 0.25 fs. The simulation procedure (see Table 2) was designed to drive the system from dispersed oligomers to condensed gel networks, following four sequential steps: first, energy minimization was conducted to eliminate unreasonable configurations; subsequently, NVT and NPT ensemble equilibration were carried out at 300 K to stabilize the initial structure and adjust density; then, the system was rapidly heated to 2000 K under NVT ensemble to provide kinetic energy for overcoming reaction barriers, followed by sufficient relaxation at this temperature to facilitate polymerization; finally, rapid quenching to 300 K and subsequent equilibration were performed to allow gel structure reorganization and stabilization.

Table 2. Simulation Process.

Step Number	Simulation Process	Ensemble Type	Temperature Condition	Time Parameter
①	Energy Minimization	—	—	—
②	Initial System Equilibration	NVT	300 K	100 ps
③	Density Equilibration	NPT	300 K	300 ps
④	Heating Process	NVT	300 K rise to 2000 K (17 K/ps)	100 ps
⑤	High-Temperature Relaxation	NVT	2000 K	1000 ps
⑥	Quenching Process	NVT	2000 K drop to 300 K (5 K/ps)	340 ps
⑦	Room-Temperature Relaxation	NVT	300 K	300 ps

This procedure balances simulation efficiency and physical realism: high-temperature relaxation accelerates the reaction, while subsequent quenching and low-temperature equilibration facilitate the formation of thermodynamically more stable structures. When the system reaches a local energy minimum state, its structural characteristics remain stable [31]. Though the limited model size affects long-range structural statistics, this approach is sufficient to capture the key dynamic processes of N-A-S-H gel formation (e.g., ion association, chemical bond formation, and local structural relaxation). Through the above procedure, well-equilibrated N-A-S-H gel structures were obtained, as illustrated in Figure 1c–e.

2.1.3. Structural Analysis and Characterization Methods

Quantitative analysis and visualization of the final equilibrated trajectories of the systems were performed using OVITO software 3.12.1, including specifically: (1) evaluating the polymerization degree by calculating the oxygen coordination numbers of individual Si and Al atoms to statistically analyze the distributions of Si^0–Si^4 and Al^0–Al^6 species; (2) extracting all Si–O and Al–O bond lengths as well as Si–O–Si, Al–O–Al and Si–O–Al bond angles, followed by analyzing their statistical distributions and average values; (3) classifying Q^n species according to the number of bridging oxygens (n = 0–4) bonded to each Si atom and calculating their proportions to characterize the network connectivity; and (4) scanning the systems using probe spheres with a radius of 1.0 Å (approximately equal to the kinetic diameter of water molecules) via the grid method, identifying and counting the volume of connected cavities accessible to the probes; the ratio of this volume to the

 https://doi.org/10.3390/ma19020246

total system volume is defined as nanoporosity, which is used to evaluate the compactness of the network structure.

2.2. Experimental Scheme

2.2.1. Raw Materials and Preparation of AAMs

Low-calcium fly ash (FA, CaO content < 10 wt%, Jinrun New Materials Co., Ltd., Zhengzhou, China) with a molar ratio of reactive SiO_2 to Al_2O_3 of approximately 1:1 was employed as the primary aluminosilicate precursor in the experiments. Waste glass powder (GP, passed through a 600-mesh sieve, Zhuangwei New Materials Co., Ltd., Zhengzhou, China) and liquid sodium silicate (LSS, modulus 3.3, solid content $\geq$ 35.5%, Yourui Refractory Materials Co., Ltd., Jiashan County, China) were used as supplementary silica sources to precisely adjust the Si/Al ratio of the system. Sodium hydroxide pellets were adopted to provide additional alkalinity and regulate the modulus of sodium silicate. The detailed chemical compositions of fly ash and glass powder are presented in Table 3.

Table 3. Material Composition Table.

Chemical Components (%)	SiO_2	Al_2O_3	CaO	Na_2O	MgO	Fe_2O_3	Others
Fly Ash	45.10	36.80	4.50	0	0	0.85	12.75
Glass Powder	71.58	0.81	4.41	13.8	3.94	0	5.46

To ensure comparability with the simulations, four mixtures with molar silica-alumina (Si/Al) ratios of 1.0, 1.5, 1.8 and 2.0 were designed. The mix proportion design followed three principles: (1) maintaining a constant Na_2O content (8 wt% of the total solid mass) to control the alkalinity, with other oxides (e.g., CaO, MgO, Fe_2O_3) excluded from this ratio calculation; (2) fixing the water-to-binder ratio at 0.4 to ensure workability and comparability; (3) adjusting the proportions of FA, GP and sodium silicate solution to achieve the target Si/Al ratios. The final Si/Al ratio of each system was calculated based on the total molar amounts of reactive SiO_2 and Al_2O_3 contributed by all solid raw materials and sodium silicate solution, with the detailed mix proportions given in Table 4.

Table 4. Mix Proportion Design of Experimental Samples (Parts by Mass).

Sample ID	Target Si/Al	FA	GP	LSS	NaOH	Add. Water	Calculated Si/Al
E-1.0	1	500	0	0	51.6	150	1.02
E-1.5	1.5	500	0	367	11.2	95	1.51
E-1.8	1.8	500	52	469	0	70	1.79
E-2.0	2	500	112	469	0	60	2.03

2.2.2. Sample Preparation and Testing Methods

Raw materials were weighed according to the mix proportions specified in Table 4. Sodium hydroxide (NaOH) was first dissolved in the calculated amount of water, and then mixed with sodium silicate solution to prepare the alkaline activator. Fly ash (FA) and glass powder (GP) were dry-mixed for 1 min, followed by the addition of the alkaline activator and mixing in a planetary mixer for 5 min to form a homogeneous paste. The paste was poured into molds with a diameter of 50 mm and a height of 50 mm, and vibrated to achieve full compaction. The samples, together with the molds, were sealed and cured in an oven at 60 °C for 24 h (elevating the temperature to 60 °C can significantly accelerate the reaction rate of fly ash without compromising the mechanical properties of the final reaction products [25]). After demolding, the samples were transferred to a standard curing

https://doi.org/10.3390/ma19020246

room (20 ± 2 °C, RH > 95%) for further curing until the ages of 1 d, 3 d, 7 d and 28 d. Characterization tests were performed at the corresponding ages, with the detailed process flow shown in Figure 2.

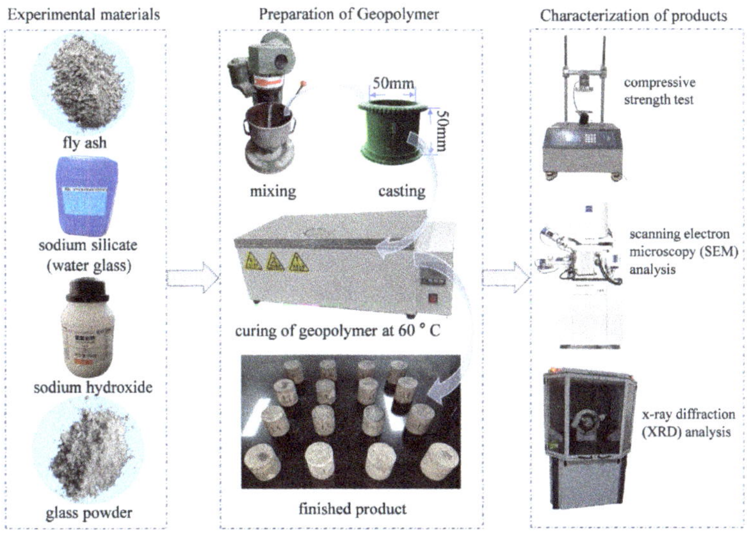

Figure 2. Experimental materials and methods.

The specific characterization methods for the reaction products are as follows:

X-ray Diffraction (XRD) Analysis: Ground powder samples were tested using an X-ray diffractometer (D8 ADVANCE, Bruker AXS GmbH, Fitchburg, WI, USA). The scanning range was 10–90° (2θ) with a step size of 0.02° and a scanning rate of 5°/min. Semi-quantitative phase analysis was conducted via the Rietveld full-spectrum fitting method.

Scanning Electron Microscopy (SEM): The morphology of samples was analyzed using a scanning electron microscope (EVO MA15, Carl Zeiss AG, Oberkochen, Germany) at an accelerating voltage of 15 kV. Prior to testing, the samples were gold-sputtered with a vacuum sputter coater, and a 5–10 nm thick gold film was deposited to enhance surface conductivity and prevent charge accumulation. The sample stage was kept stable during testing to ensure clear imaging and reliable results.

Unconfined Compressive Strength (UCS) Test: In accordance with ASTM C109 [32] and JTG 3441-2024 [33], the compressive strength of specimens was tested using an unconfined compressive strength testing machine (TKA-WXY-6F, Nanjing Tekao Technology Co., Ltd., Nanjing, China). The tests were performed at 25 °C with a loading rate of 5 mm/min. For each mix proportion, five parallel specimens were tested, and the results were averaged.

3. Results and Discussion

3.1. Microstructural Evolution Law Revealed by Molecular Dynamics Simulation

3.1.1. Gelation Kinetics and the Dominant Role of Aluminum Coordination

The simulation successfully reproduced the dynamic process from discrete PS/PSS oligomers to a continuous gel network, as illustrated in Figure 3. In the construction of the N-A-S-H gel model, all initial aluminosilicate species existed as PS and PSS oligomers. The essence of their polymerization reaction lies in the process where PS and PSS oligomers form bridging oxygen (Q^n) crosslinking sites via dehydration of OH and H groups, accompanied by water molecule release. As the reaction proceeded, these structural units underwent

 https://doi.org/10.3390/ma19020246

continuous condensation to form the aluminosilicate network skeleton. When the reaction proceeded to 2000 ps, the aluminosilicate chains exhibited low dispersion in the system with Si/Al = 1.0, whereas the chain dispersion increased significantly under the condition of Si/Al = 2.0.

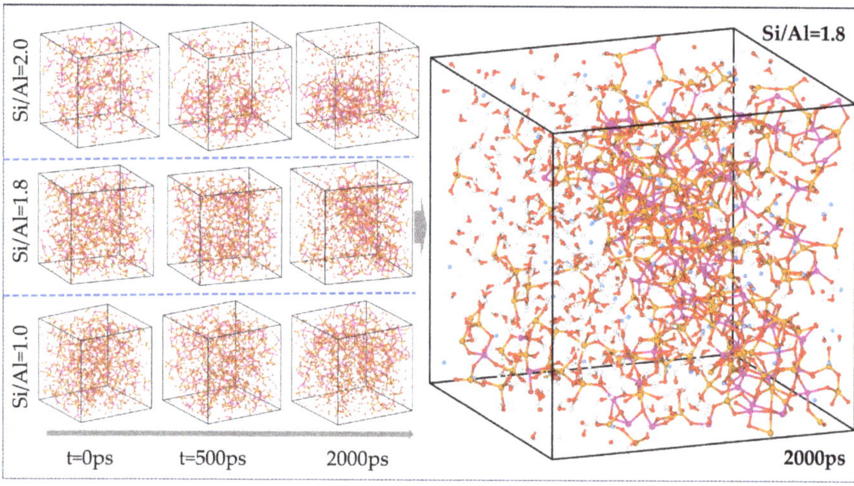

Figure 3. Structural evolution of N A S H gel for Si/Al ratios.

By tracking the decay in the number of non-bridging oxygens (Si–OH and Al–OH) and the content variation in newly formed Q^n species (Figures 4 and 5), it was found that the consumption rate of Al–OH was faster than that of Si–OH under all Si/Al conditions, which kinetically confirms the higher reactivity of Al species. This is consistent with the theoretical understanding that the dissociation energy of Al–O bonds is generally lower than that of Si–O bonds, and also explains the common phenomenon of preferential Al dissolution observed in experiments. As the simulation proceeded, non-bridging oxygens were continuously converted into Q^n species with higher polymerization degrees (Si^n and Al^n, n = 1–4): Si^1 appeared first, followed by the sequential formation of Si^2, Si^3 and Si^4, indicating that the reaction follows a stepwise polymerization pathway from monomers to oligomers and then to large clusters, which is consistent with the actual geopolymerization process [34]. At the end of the simulation, most non-bridging oxygens were consumed, with only a small amount remaining due to intrinsic topological defects in the amorphous network (e.g., broken bonds and under-coordinated sites). These defects render the network more susceptible to OH^- attack and promote the preferential formation of Si^3 during depolymerization, resulting in the highest proportion of Si^3 in the end. The evolution of the aluminooxygen network was similar to that of the silicooxygen network; Al–OH was almost completely consumed eventually, but a considerable proportion of pentacoordinate aluminum (Al^5) and a small amount of hexacoordinate aluminum (Al^6) were formed during the process [35]—the defect sites in the amorphous network provide a stable environment for Al^5, leading to its final proportion exceeding 50%.

Coordination analysis of the final gel structure reveals a striking feature that deviates from conventional understanding: pentacoordinate aluminum (Al^5) dominates as the primary Al species across all four Si/Al systems, with its proportion stably exceeding 50% of the total aluminum content (Figure 6). In contrast, the classic tetracoordinate aluminum (Al^4) accounts for approximately 30–40%, along with a small amount of hexacoordinate aluminum (Al^6). This result differs from numerous early simulations based on ideal tetrahe-

https://doi.org/10.3390/ma19020246

dral models and some NMR studies, but is consistent with recent advanced characterization findings under high-alkalinity and non-equilibrium synthesis conditions (e.g., the works by Wang Z et al. [36–38]) as well as the trends of several latest reactive force field simulations [11,39]. We propose that in the dynamic, water-rich strong alkaline environment depicted by ReaxFF simulations, Al^5 is not a "defect" but rather a key intermediate stable state or charge regulator. Its formation stems from two aspects: on the one hand, the strong affinity of Al for hydroxyl ligands under high pH values; on the other hand, it may serve to balance local structural stress and negative charges more effectively, especially when Na^+ exhibits uneven spatial distribution. Al^5 sites bear more hydroxyl groups and may act as an "active reservoir" for condensation reactions, facilitating the rapid growth of the gel network.

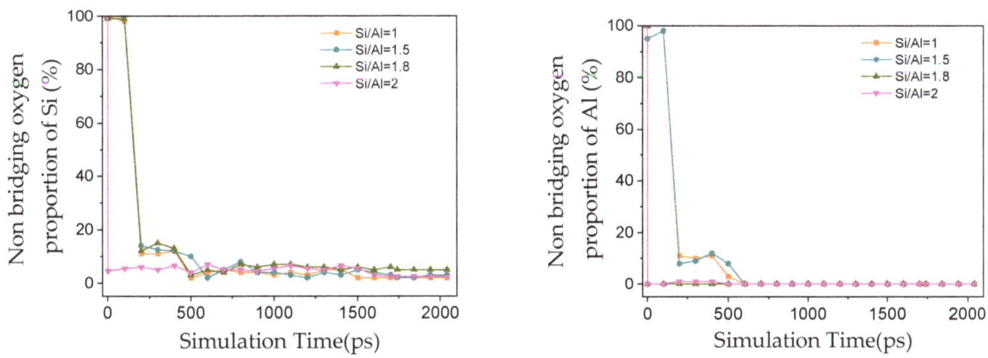

Figure 4. Evolution of non-bridging oxygen proportion for silicon and aluminum species with reaction time under different Si/Al ratios.

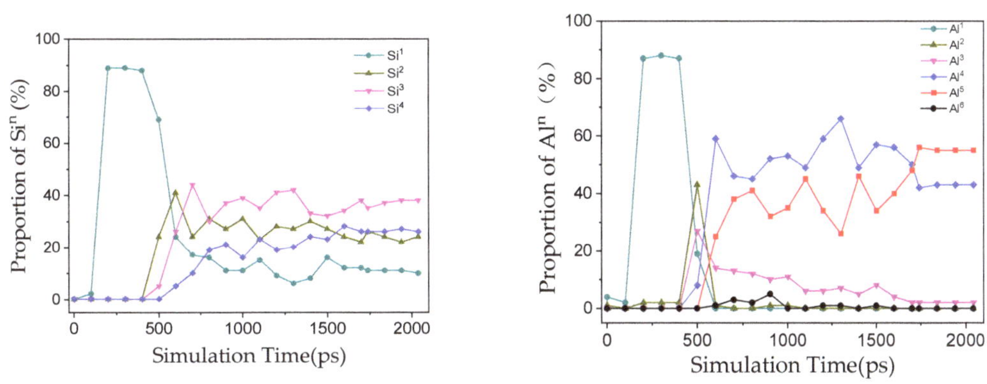

Figure 5. Evolution of Q^n species during the condensation reaction (Si/Al = 1.0).

3.1.2. Network Topology and Bonding Environment

The connection state of silicon-oxygen tetrahedra (Q^n distribution) is a key indicator for evaluating network polymerization degree. Specifically, Q^1 represents terminal sites, Q^2 denotes middle groups of chain/cyclic structures, Q^3 refers to branched structures, and Q^4 indicates fully cross-linked structures [40,41]; the overall structural composition can be inferred from their distribution proportions. As shown in Figure 6, a considerable proportion of high-connectivity species (Q^3 + Q^4, total 53–64%) were generated in all systems, confirming the formation of three-dimensional (3D) networks. Notably, as the Si/Al ratio increased from 1.0 to 2.0, the ratio of ($Q^3 + Q^4$)/($Q^1 + Q^2$) exhibited a decreasing

 https://doi.org/10.3390/ma19020246

trend. This indicates that lower Si/Al ratios (1.0, 1.5) are more conducive to the formation of highly cross-linked networks, whereas higher Si/Al ratios (1.8, 2.0) may lead the network connectivity topology to tend toward lower dimensionality or more chain-like structures. This is consistent with the chemical intuition that an increase in Si/Al ratio results in a relative reduction in Al sites available for cross-linking.

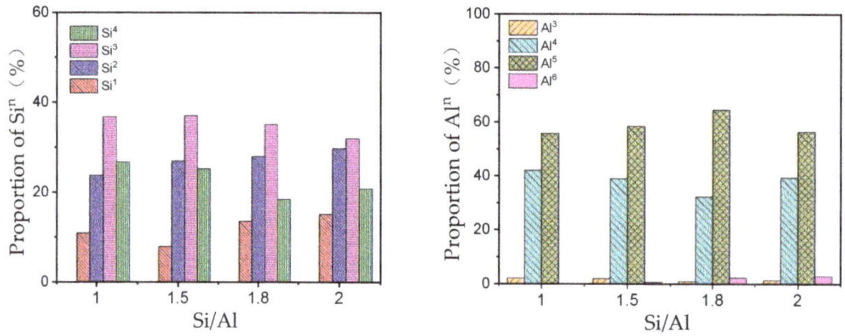

Figure 6. Distribution of Si^n and Al^n sites in N-A-S-H gel models at different Si/Al ratios.

Bond angle distribution provides more refined structural information (Figure 7). The Si-O-Si bond angle exhibited a unimodal distribution with a peak at approximately 140° and a narrow distribution range, reflecting the intrinsic rigidity of $[SiO_4]^{4-}$ tetrahedral connections. In contrast, the Al-O-Al bond angle showed a unique bimodal distribution (~100° and ~150°), which directly reflects the significant local structural distortion caused by aluminum coordination polymorphism (coexistence of Al^4 and Al^5/Al^6). This flexibility or "disorder degree" of bond angles is an important feature that distinguishes N-A-S-H gels from crystalline zeolites, and may also serve as potential weak points for stress concentration.

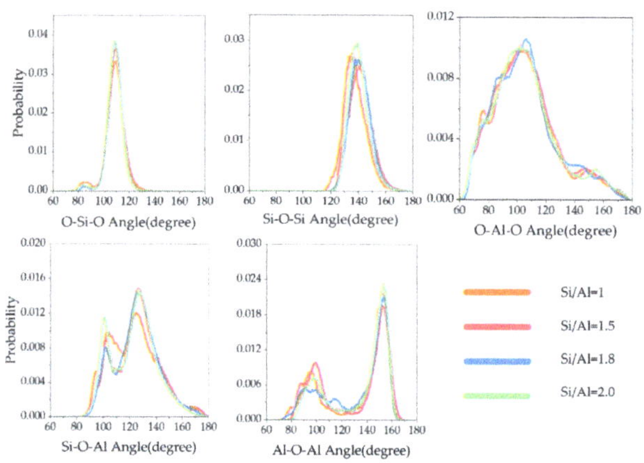

Figure 7. Bond angle distribution in the N-A-S-H Gel model.

Bond length distribution reveals the fundamental linkage characteristics of the gel network (Figure 8). The simulated average Si–O bond length (1.64 Å) is in good agreement with reported literature values, while the average Al–O bond length (1.84 Å) is systematically higher [12,42]. This difference mainly stems from the high proportion of pentacoordinate aluminum (Al^5) predicted by this model; the Al–O bond length of Al^5 is inherently longer than that of common tetracoordinate aluminum (Al^4), thus elevating the overall average value. In addition, as the Si/Al ratio increases from 1.0 to 2.0, both Si–O

https://doi.org/10.3390/ma19020246

and Al–O bond length distributions show a slight right shift, indicating a corresponding increase in average bond length. This trend is consistent with existing studies and explains the macroscale mechanism by which increasing Si/Al ratio reduces the elastic modulus of N-A-S-H gels from an atomic perspective—longer bond lengths generally correspond to weaker bonding strength and lower network stiffness.

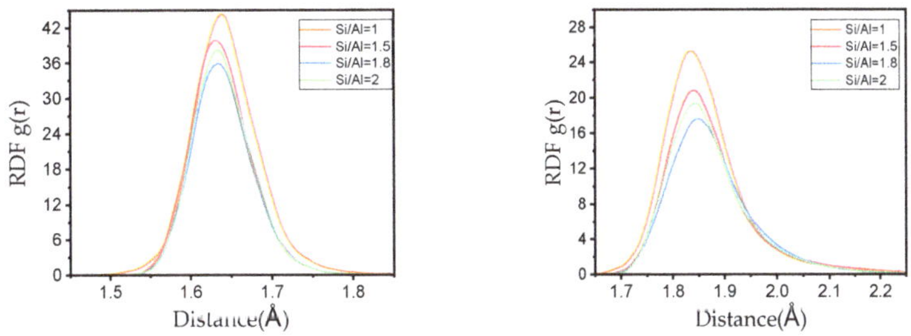

Figure 8. Bond length distribution of Si-O (**left**) and Al-O (**right**) in the N-A-S-H Gel model.

3.1.3. Non-Monotonic Nanoporosity Variation and Optimal Si/Al Ratio

The calculated nanoporosity (probe radius: 1.0 Å) exhibited a clear non-monotonic trend of first decreasing and then increasing with Si/Al ratio (Figure 1e). It decreased gradually from 49.3% at Si/Al = 1.0, reached a minimum of 35.2% at Si/Al = 1.8, and then rose back to 42.1% at Si/Al = 2.0, as shown in Figure 9. The existence of this inflection point is crucial, indicating that the atomic packing density of the gel network does not improve monotonically with increasing Si content but has an optimal stoichiometric balance point (~1.8).

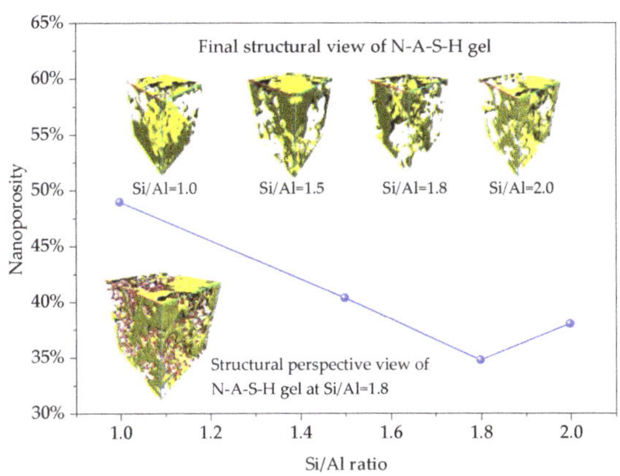

Figure 9. Formation and distribution of nanopores in N-A-S-H gel models at different Si/Al ratios, where yellow indicates pore regions.

The mechanism underlying this phenomenon is explained as follows: In the low Si/Al region (≤1.8), increasing Si content provides more $[SiO_4]^{4-}$ tetrahedral units for skeleton construction, which combine with sufficient Al cross-linking sites to form a more complete and compact network, leading to decreased porosity. At Si/Al = 1.8, the Si/Al ratio achieves optimal synergy, enabling sufficient network cross-linking, robust skeleton, high Na^+ charge compensation efficiency, the densest structure and optimal mechanical properties. When the Si/Al ratio further increases (>1.8), the relative shortage of Al atoms available for effective cross-linking causes excessive Si to tend to form longer linear

 https://doi.org/10.3390/ma19020246

Si–O–Si segments with low packing efficiency. Meanwhile, the high proportion of Al^5/Al^6 requires more isolated Na^+ compensation, and these hydrated Na^+ ions occupy extra space. Together, these factors result in increased nanoscale free volume and thus porosity rebound.

3.2. Experimental Validation: Correlation Between Macroscopic Properties and Microstructure

3.2.1. Mechanical Property Response to Si/Al Ratio

The experimentally measured unconfined compressive strength (UCS) results (Figure 10) provide strong macroscopic evidence for the simulation predictions. The specimens exhibited predominantly brittle fracture, and the strength variation curve with Si/Al ratio exhibits a distinct peak characteristic. Statistical analysis showed that both curing age and Si/Al ratio had an extremely significant effect on UCS ($p < 0.0001$), and the strength values of the Si/Al = 1.8 group were significantly higher than those of other groups. Specimens with Si/Al = 1.8 achieved the highest strength at both 7 d and 28 d (specifically: average value 9.4 MPa, standard deviation 0.283 MPa, 95% confidence interval [9.05, 9.75] MPa; and average value 15.3 MPa, standard deviation 0.539 MPa, 95% confidence interval [14.64, 15.98] MPa). Statistical analysis (one-way ANOVA, $p < 0.05$) indicated that the strength values of this group were significantly higher than those of other groups. For specimens with Si/Al = 1.0, insufficient silicon source resulted in a limited total amount of cementitious phases formed by the reaction, leading to a loose structure and the lowest strength. Meanwhile, the strength of specimens with Si/Al = 2.0 decreased compared to the Si/Al = 1.8 group, which is fully consistent with the simulation predictions of increased porosity and reduced structural compactness at this ratio. The inverse correlation between macroscopic strength and simulated nanoporosity establishes a key bridge from atomic-scale structure to macroscopic properties.

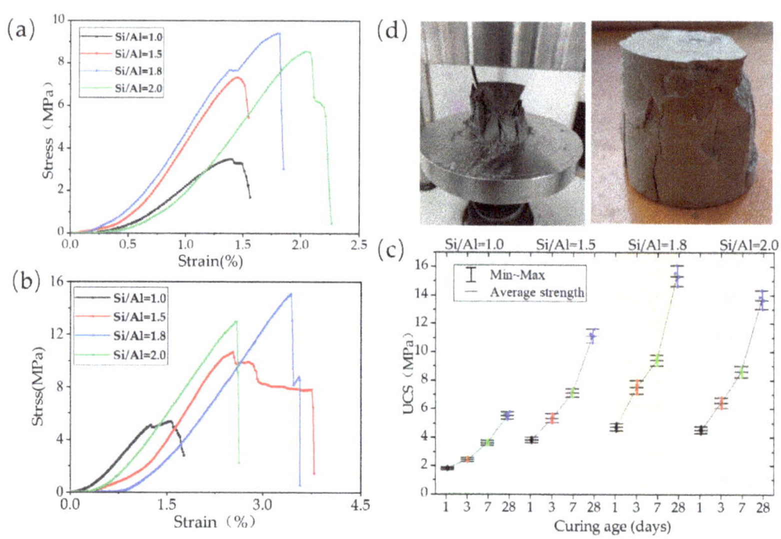

Figure 10. Evolution of mechanical properties: (**a**,**b**) stress–strain curves at 7 and 28 days of curing, respectively; (**c**) development of compressive strength over different curing ages (1, 3, 7, and 28 days). (**d**) Representing the failure morphology of the specimen.

3.2.2. Evidence of Phase Evolution and Microscopic Morphology

X-ray diffraction (XRD) combined with Rietveld refinement was employed for quantitative phase analysis of alkali-activated cementitious materials at 7 d and 28 d curing ages. The results (Figure 11c) showed that as the Si/Al ratio increased from 1.0 to 1.8,

https://doi.org/10.3390/ma19020246

the contents of quartz and mullite in the raw materials decreased continuously, while the content of amorphous N-A-S-H gel product increased correspondingly (The parts enclosed by the blue dashed boxes in Figure 11a,b), indicating a gradual improvement in reaction degree. At Si/Al = 1.8, the residual raw materials were minimized and the gel product was maximized, corresponding to the highest reactivity and conversion rate. When the Si/Al ratio further increased to 2.0, the reaction degree decreased slightly, which may be related to the excess silicon altering the liquid-phase chemical equilibrium and affecting precursor dissolution and condensation kinetics.

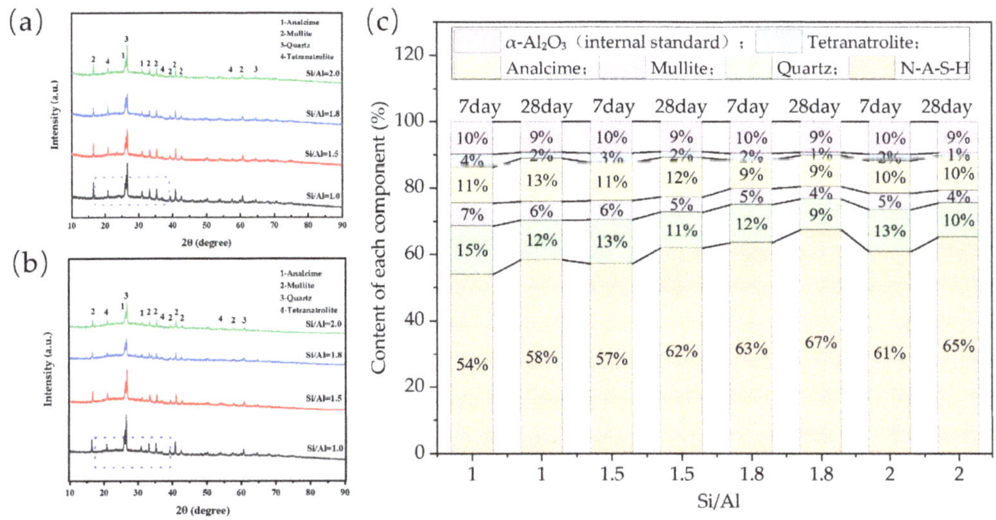

Figure 11. XRD patterns and phase content of samples with different Si/Al ratios at 7 and 28 days. (**a**) 7 d XRD patterns of specimens with different Si/Al ratios; (**b**) 28 d XRD patterns of specimens with different Si/Al ratios; (**c**) Quantitative analysis results.

XRD data further revealed the dynamic phase evolution during the reaction: in the alkaline environment, crystalline phases such as quartz and mullite continuously dissolved, releasing reactive aluminosilicate species that formed amorphous N-A-S-H gel via condensation. This gel gradually encapsulated unreacted particles. As the reaction proceeded, metastable Tetranatrolite precipitated from part of the gel or liquid phase; in the long-term reaction, it gradually dissolved and transformed into thermodynamically more stable analcime crystals via the Ostwald ripening mechanism, leading to a decrease in the content of the former and a significant increase in the latter. Meanwhile, the three-dimensional (3D) amorphous gel network formed by the condensation of aluminosilicate oligomers continued to develop and densify. Eventually, the system evolved into a composite structure consisting of amorphous N-A-S-H gel and stable zeolite crystals (predominantly analcime). The formation of crystalline phases helps improve the long-term chemical stability and compactness of the system, and is one of the key factors for geopolymers to achieve long-term strength, thus explaining the highest strength of specimens with Si/Al = 1.8 at the mesoscopic scale.

Scanning electron microscopy (SEM) observations (Figure 12) provided more intuitive microstructural evidence. In samples with Si/Al = 1.0, a large number of unreacted fly ash spherical particles were dispersed in the loose gel matrix. For samples with Si/Al = 1.5 and 1.8, the number of unreacted particles decreased, and the gel phase became more continuous and dense. Especially in the sample with Si/Al = 1.8, the gel exhibited a uniform and dense

https://doi.org/10.3390/ma19020246

morphology with few pores. In contrast, although the number of unreacted particles was also small in the sample with Si/Al = 2.0, some microcracks and finer pores appeared in the gel phase, resulting in slightly poor structural uniformity. These observations directly support the conclusion from simulations that the optimal dense structure occurs around Si/Al = 1.8.

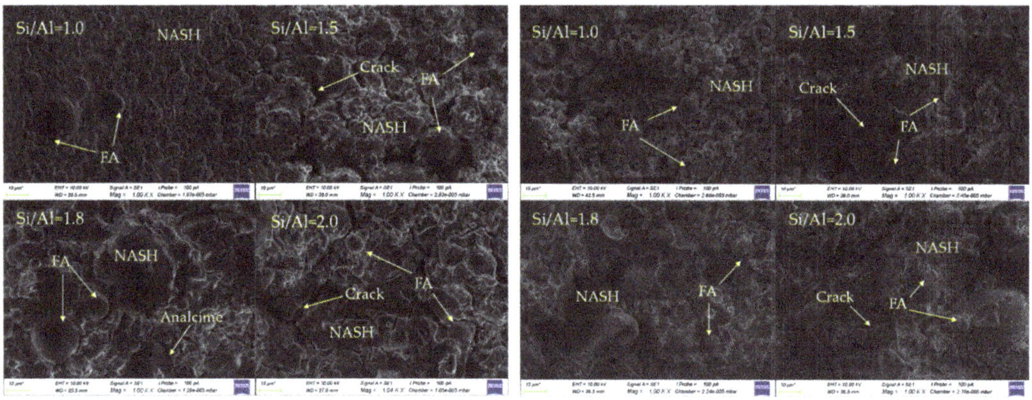

(a) SEM (7day) (b) SEM (28day)

Figure 12. SEM Images of Geopolymer Products with Different Si/Al Molar Ratios.

3.3. Comprehensive Discussion: Regulation Mechanism of Si/Al Ratio and Structure-Property Relationship

Based on the in-depth mutual verification of the aforementioned simulation and experimental results, we propose an integrated mechanism for the regulation of N-A-S-H gel structure and properties by Si/Al ratio:

1. Al-rich region (Si/Al < 1.8): Sufficient Al species can rapidly form cross-linking sites, promoting the three-dimensionalization of the network. However, the silicon skeleton is relatively weak, and a large amount of Al may exist in high-coordination states (Al^5, Al^6), requiring more Na^+ for charge compensation. Insufficient compensation or large steric hindrance tends to form local loose structural regions. The final structure has a moderate degree of cross-linking but inherently high porosity, corresponding to moderate macroscopic strength.
2. Optimal region (Si/Al $\approx$ 1.8): The Si/Al ratio achieves an optimal match between thermodynamics and kinetics. Al provides moderate cross-linking sites, while Si forms a robust and moderately branched skeleton. Al^5 exists as an effective charge buffer and reaction intermediate, achieving efficient synergy with Na^+. Under this condition, the condensation reaction degree is high, forming an ideal gel network with high cross-linking, low porosity, and high continuity, corresponding to the peak macroscopic strength.
3. Si-rich region (Si/Al > 1.8): Al cross-linking sites are relatively scarce, and Si tends to polymerize into long-chain structures, leading to a decrease in network cross-linking density. Limited Al mostly exists in high-coordination states to balance the system charge, exacerbating structural disorder. Long-chain silicates have poor close packing ability, and the isolation effect of Na^+ is more significant. Together, these factors result in a significant rebound in nanoporosity and a decrease in the overall structural stiffness and strength of the gel.

https://doi.org/10.3390/ma19020246

The core of this mechanism lies in the synergy and competition between "aluminum coordination polymorphism" and "silicate polymerization degree." By regulating the balance of these two factors, the Si/Al ratio ultimately determines the nanoscale topology and pore characteristics of the gel network, thereby controlling the macroscopic mechanical properties. In this study, the key role of Al^5, the non-monotonic variation in nanoporosity, and its optimal value at Si/Al = 1.8 collectively constitute the microscopic evidence chain for this regulation mechanism.

Although this study reveals the regulation mechanism of the Si/Al ratio on N-A-S-H gels, its in-depth understanding is still limited by the size and time scale of the computational model, force field accuracy, simplified chemical composition (Ca, Fe and other elements not considered), and the coupling effect of experimental variables. Future research can develop cross-scale computational methods combined with multi-variable collaborative experiments to achieve more accurate prediction and design of material properties in systems closer to real conditions.

4. Conclusions

In this study, by integrating reactive molecular dynamics simulations and systematic experimental studies, and spanning atomic–nano–micro–macro scales, we profoundly revealed the regulation laws and mechanisms of the Si/Al ratio on the structure and properties of alkali-activated N-A-S-H gels, and obtained the following main conclusions:

1. Revealed the dominant role of pentacoordinate aluminum: In the dynamic condensation environment, pentacoordinate aluminum (Al^5) is the main existing form of aluminum in N-A-S-H gels (>50%). It may play a key role in charge balance and reaction promotion, which updates the understanding of the local structure of gels.
2. Discovered the optimal ratio for structural densification: Molecular simulations clearly show that the nanoporosity of the gel reaches a minimum (35.2%) at Si/Al $\approx$ 1.8, indicating the highest atomic packing efficiency and optimal network compactness at this ratio.
3. Established a cross-scale structure-property relationship: Experiments confirmed that the specimen with Si/Al = 1.8 exhibits the highest reaction degree, the most uniform and dense microscopic morphology, and the highest macroscopic compressive strength (15.3 MPa at 28 d), which perfectly verifies the simulation predictions and establishes a strong correlation between nanoporosity and macroscopic strength.
4. Clarified the regulation mechanism of Si/Al ratio: This study proposes that the Si/Al ratio regulates the gel structure by regulating the balance between "Al cross-linking capacity" and "Si skeleton construction capacity". Deviation from the optimal ratio will lead to insufficient cross-linking or excessive skeleton polymerization, thereby causing an increase in nanoporosity and performance degradation.

This study not only provides atomic-scale insights into understanding the intrinsic origin of performance of alkali-activated materials, but more importantly, the revealed optimal Si/Al ratio (~1.8) and related structural characteristic parameters (low porosity, appropriate Al^5 proportion) can offer theoretical guidance and quantitative basis for the rational design and precise regulation of alkali-activated cementitious materials targeting high-performance requirements.

Author Contributions: Conceptualization, M.H. and J.C. (Jiayun Chen); Methodology, M.H., J.C. (Jiayun Chen) and B.X.; Investigation, M.H. and J.C. (Jiayun Chen); Resources, M.H., J.C. (Jiayun Chen), B.X. and J.C. (Jiejin Chen); Data curation, J.C. (Jiayun Chen); Supervision, M.H. and J.C. (Jiejin Chen); Writing—original draft, M.H. and J.C. (Jiayun Chen); Writing—review &

 https://doi.org/10.3390/ma19020246

editing, M.H., J.C. (Jiayun Chen), B.X. and J.C. (Jiejin Chen). All authors have read and agreed to the published version of the manuscript.

Funding: This work is funded by the National Natural Science Foundation of China Youth Program No. 51908065, the Science and Technology Progress and Innovation Program Project of the Hunan Provincial Department of Transportation No. 201938, the Scientific Research Project of the Hunan Provincial Department of Education No. 20B010, and the Open Fund for Changsha University of Science and Technology Bridge Engineering No. 15KC03.

Data Availability Statement: The original contributions presented in this study are included in the article. Further inquiries can be directed to the corresponding authors.

Conflicts of Interest: Author Min Hu is employed by Changsha University of Science and Technology. Author Jiayun Chen is employed by Changsha University of Science and Technology. Author Bo Xia is employed by Changsha University of Science and Technology. Author Jiejin Chen is employed by Changsha University of Science and Technology. The authors declare that the research was conducted in the absence of any commercial or financial relationships that could be construed as a potential conflict of interest.

References

1. Qaidi, S.; Najm, H.M.; Abed, S.M.; Özkılıç, Y.O.; Al Dughaishi, H.; Alosta, M.; Sabri, M.M.S.; Alkhatib, F.; Milad, A. Concrete containing waste glass as an environmentally friendly aggregate: A review on fresh and mechanical characteristics. *Materials* **2022**, *15*, 6222. [CrossRef] [PubMed]
2. Segura, I.P.; Ranjbar, N.; Damø, A.J.; Jensen, L.S.; Canut, M.; Jensen, P.A. A review: Alkali-activated cement and concrete production technologies available in the industry. *Heliyon* **2023**, *9*, e15718. [CrossRef] [PubMed]
3. Qader, D.N.; Jamil, A.S.; Bahrami, A.; Ali, M.; Arunachalam, K.P. A systematic review of metakaolin-based alkali-activated and geopolymer concrete: A step toward green concrete. *Rev. Adv. Mater. Sci.* **2025**, *64*, e01958. [CrossRef]
4. Wang, H.; Zhao, X.; Ma, B.; Wang, T.; Su, L.; Zhou, B.; Lin, Y. Determination of gel products in alkali-activated fly ash-based composites incorporating inorganic calcium additives. *Adv. Mater. Sci. Eng.* **2022**, *2022*, 7476671. [CrossRef]
5. Sun, B.; Ye, G.; De Schutter, G. A review: Reaction mechanism and strength of slag and fly ash-based alkali-activated materials. *Constr. Build. Mater.* **2022**, *326*, 126843. [CrossRef]
6. Chen, Y.; de Lima, L.M.; Li, Z.; Ma, B.; Lothenbach, B.; Yin, S.; Yu, Q.; Ye, G. Synthesis, solubility and thermodynamic properties of N-A-S-H gels with various target Si/Al ratios. *Cem. Concr. Res.* **2024**, *180*, 107484. [CrossRef]
7. Duxson, P.; Provis, J.L.; Lukey, G.C.; Mallicoat, S.W.; Kriven, W.M.; van Deventer, J.S.J. Understanding the relationship between geopolymer composition, microstructure and mechanical properties. *Colloids Surf. A Physicochem. Eng. Asp.* **2005**, *269*, 47–58. [CrossRef]
8. Wan, Q.; Rao, F.; Song, S.; García, R.E.; Estrella, R.M.; Patiño, C.L.; Zhang, Y. Geopolymerization reaction, microstructure and simulation of metakaolin-based geopolymers at extended Si/Al ratios. *Cem. Concr. Compos.* **2017**, *79*, 45–52. [CrossRef]
9. Bumanis, G.; Vitola, L.; Bajare, D.; Dembovska, L.; Pundiene, I. Impact of reactive SiO_2/Al_2O_3 ratio in precursor on durability of porous alkali activated materials. *Ceram. Int.* **2017**, *43*, 5471–5477. [CrossRef]
10. Zhou, X.; Chen, Y.; Dong, S.; Li, H. Geopolymerization kinetics of steel slag activated gasification coal fly ash: A case study for amorphous-rich slags. *J. Clean. Prod.* **2022**, *379*, 134671. [CrossRef]
11. Chen, Y.; Dolado, J.S.; Li, Z.; Yin, S.; Yu, Q.; Kostiuchenko, A.; Ye, G. A molecular dynamics study of N–A–S–H gel with various Si/Al ratios. *J. Am. Ceram. Soc.* **2022**, *105*, 6462–6474. [CrossRef]
12. Wang, R.; Wang, J.S.; Dong, T.; Ouyang, G. Structural and mechanical properties of geopolymers made of aluminosilicate powder with different SiO_2/Al_2O_3 ratio: Molecular dynamics simulation and microstructural experimental study. *Constr. Build. Mater.* **2020**, *240*, 117935. [CrossRef]
13. Xu, W.; Tang, Z.; Xie, Y.; Long, G.; Kai, M.; Zhang, Z.; Bu, M.; Zaland, S. Understanding the impact of synthesis parameters on the pore structure properties of fly ash-based geopolymers. *Constr. Build. Mater.* **2024**, *411*, 134640. [CrossRef]
14. Xu, L.Y.; Alrefaei, Y.; Wang, Y.S.; Dai, J.G. Recent advances in molecular dynamics simulation of the N-A-S-H geopolymer system: Modeling, structural analysis, and dynamics. *Constr. Build. Mater.* **2021**, *276*, 122196. [CrossRef]
15. Li, L.; Wei, Y.; Feng, Q.; Liu, F.; Liu, B.; Pu, B. A review: Progress in molecular dynamics simulation of Portland cement (geopolymer)—Based composites and the interface between these matrices and reinforced material. *Buildings* **2023**, *13*, 1875. [CrossRef]

https://doi.org/10.3390/ma19020246

16. Yang, H.L.; Ma, S.Q.; Zhao, S.J.; Wang, Q.; Liu, X.; He, P.; Jia, D.; Colombo, P.; Zhou, Y. Mechanistic understanding of geopolymerization at the initial stage: Ab initio molecular dynamics simulations. *J. Am. Ceram. Soc.* **2023**, *106*, 4425–4442. [CrossRef]
17. Guo, F.; Chen, J.; Tang, Q.; Sun, M.; Feng, H.; Gao, H.; Li, M.; Lu, S. Molecular dynamics simulation of the initial stage induction of alkali-activated aluminosilicate minerals. *RSC Adv.* **2024**, *14*, 13972–13983. [CrossRef]
18. Luo, X.; Tian, X.; Wu, J.; Yang, X.; Liu, Z.; Jiao, Z.; Peng, H. Molecular simulations of the initial stage's induction and formation process of N-A-S-H Gel based on NaOH-activated Metakaolin. *J. Non-Cryst. Solids* **2024**, *626*, 122804. [CrossRef]
19. Liu, Y.L.; Ren, S.Y.; Wang, D.H.; Yang, D.-W.; Kai, M.-F.; Guo, D. Molecular structure and thermal conductivity of hydrated sodium aluminosilicate (N-A-S-H) gel under different Si/Al ratios and temperatures: A molecular dynamics analysis. *AIMS Mater. Sci.* **2025**, *12*, 258–277. [CrossRef]
20. Guan, X.; Jiang, L.; Fan, D.; Hernandez, A.G.; Li, B.; Do, H. Molecular simulations of the structure-property relationships of N-A-S-H gels. *Constr. Build. Mater.* **2022**, *329*, 127166. [CrossRef]
21. Noorpour, M.; Tarighat, A. Effects of Si/Al ratio on structure, modulus of elasticity, and density in N-A-S-H geopolymer: A molecular dynamics simulation based on novel macromolecular model. *J. Mol. Model.* **2021**, *27*, 342. [CrossRef] [PubMed]
22. Davidovits, J. *Geopolymer Chemistry and Applications*, 2nd ed.; Institut Géopolymère: Saint-Quentin, France, 2008.
23. Shamo, E.; Charpentier, T.; Moskura, M.; Chartier, A.; Miserque, F.; Rousselet, A. Synthesis methodology of pure N–A–S–H gels with wide range of Si/Alratios at ambient temperature. *J. Am. Ceram. Soc.* **2024**, *107*, 7537–7549. [CrossRef]
24. Li, J.; Gao, M.; Yan, W.; Yu, J. Regulation of the Si/Al ratios and Al distributions of zeolites and their impact on properties. *Chem. Sci.* **2023**, *14*, 1935–1959. [CrossRef] [PubMed]
25. Kumar, S.; Kumar, R. Mechanical activation of fly ash: Effect on reaction, structure and properties of resulting geopolymer. *Ceram. Int.* **2011**, *37*, 533–541. [CrossRef]
26. van Duin, A.C.T.; Dasgupta, S.; Lorant, F.; Goddard, W.A. ReaxFF: A Reactive Force Field for Hydrocarbons. *J. Phys. Chem. A* **2001**, *105*, 9396–9409. [CrossRef]
27. van Duin, A.C.T.; Strachan, A.; Stewman, S.; Zhang, Q.; Xu, X.; Goddard, W.A. ReaxFFSiO Reactive Force Field for Silicon and Silicon Oxide Systems. *J. Phys. Chem. A* **2003**, *107*, 3803–3811. [CrossRef]
28. Yan, Y.G.; Wang, W.S.; Li, W.; Loh, K.P.; Zhang, J. A graphene-like membrane with an ultrahigh water flux for desalination. *Nanoscale* **2017**, *9*, 18951–18958. [CrossRef]
29. Zhang, Y.; Zhang, J.; Jiang, J.; Hou, D.; Zhang, J. The effect of water molecules on the structure, dynamics, and mechanical properties of sodium aluminosilicate hydrate (NASH) gel: A molecular dynamics study. *Constr. Build. Mater.* **2018**, *193*, 491–500. [CrossRef]
30. Pitman, M.C.; van Duin, A.C.T. Dynamics of confined reactive water in smectite clay–zeolite composites. *J. Am. Chem. Soc.* **2012**, *134*, 3042–3053. [CrossRef]
31. Shahsavari, R.; Buehler, M.J.; Pellenq, R.J.; Ulm, F. First-principles study of elastic constants and interlayer interactions of complex hydrated oxides: Case study of tobermorite and jennite. *J. Am. Ceram. Soc.* **2009**, *92*, 2323–2330. [CrossRef]
32. *ASTM C109*; Standard Test Method for Compressive Strength of Hydraulic Cement Mortars. ASTM International: West Conshohocken, PA, USA, 2025.
33. *JTG 3441-2024*; Test Methods of Inorganic Binder Stabilized Materials for Highway Engineering. Ministry of Transport of the People's Republic of China: Beijing, China, 2024.
34. Duxson, P.; Fernández-Jiménez, A.; Provis, J.L.; Lukey, G.C.; Palomo, A.; van Deventer, J.S.J. Geopolymer technology: The current state of the art. *J. Mater. Sci.* **2007**, *42*, 2917–2933. [CrossRef]
35. Jin, M.; Ravi, M.; Lei, C.; Heard, C.J.; Brivio, F.; Tošner, Z.; Grajciar, L.; van Bokhoven, J.A.; Nachtigall, P. Dynamical equilibrium between Brønsted and Lewis sites in zeolites: Framework-associated octahedral aluminum. *Angew. Chem. Int. Ed.* **2023**, *62*, e202306183. [CrossRef] [PubMed]
36. Wang, Z.C.; Jiang, Y.J.; Baiker, A.; Hunger, M.; Huang, J. Promoting aromatic C–H activation through reactive Brønsted acid–base pairs on penta-coordinated Al-enriched amorphous silica–alumina. *J. Phys. Chem. Lett.* **2022**, *13*, 486–491. [CrossRef] [PubMed]
37. Wang, Z.; Jiang, Y.; Baiker, A.; Huang, J. Pentacoordinated aluminum species: New frontier for tailoring acidity-enhanced silica–alumina catalysts. *Acc. Chem. Res.* **2020**, *53*, 2648–2658. [CrossRef]
38. Wang, Z.C.; Jiang, Y.J.; Yi, X.F.; Zhou, C.; Rawal, A.; Hook, J.; Liu, Z.; Deng, F.; Zheng, A.; Hunger, M.; et al. High population and dispersion of pentacoordinated AlV species on the surface of flame-made amorphous silica-alumina. *Sci. Bull.* **2019**, *64*, 516–523. [CrossRef]
39. Zhang, W.; van Duin, A.C.T. Improvement of the ReaxFF description for functionalized hydrocarbon/water weak interactions in the condensed phase. *J. Phys. Chem. B* **2018**, *122*, 4083–4092. [CrossRef]
40. Brunet, F.; Cabane, B.; Dubois, M.; Perly, B. Sol-gel polymerization studied through silicon-29 NMR with polarization transfer. *J. Phys. Chem.* **1991**, *95*, 945–951. [CrossRef]

 https://doi.org/10.3390/ma19020246

41. Singh, P.S.; Bastow, T.; Trigg, M. Structural studies of geopolymers by ^{29}Si and ^{27}Al MAS-NMR. *J. Mater. Sci.* **2005**, *40*, 3951–3961. [CrossRef]
42. White, C.E.; Provis, J.L.; Proffen, T.; van Deventer, J.S.J. The effects of temperature on the local structure of metakaolin-based geopolymer binder: A neutron pair distribution function investigation. *J. Am. Ceram. Soc.* **2010**, *93*, 3486–3492. [CrossRef]

https://doi.org/10.3390/ma19020246

MDPI AG
Grosspeteranlage 5
4052 Basel
Switzerland
Tel.: +41 61 683 77 34

MDPI Books Editorial Office
E-mail: books@mdpi.com
www.mdpi.com/books

www.ingramcontent.com/pod-product-compliance
Lightning Source LLC
Chambersburg PA
CBHW061616210926
51850CB00009B/2339
* 9 7 8 3 7 2 5 8 6 9 6 2 6 *